T0254872

Sources and Studies in the History of Mathematics and Physical Sciences

Sources and Studies in the History of Mathematics and Physical Sciences was inaugurated as two series in 1975 with the publication in Studies of Otto Neugebauer's seminal three-volume History of Ancient Mathematical Astronomy, which remains the central history of the subject. This publication was followed the next year in Sources by Gerald Toomer's transcription, translation (from the Arabic), and commentary of Diocles on Burning Mirrors. The two series were eventually amalgamated under a single editorial board led originally by Martin Klein (d. 2009) and Gerald Toomer, respectively two of the foremost historians of modern and ancient physical science. The goal of the joint series, as of its two predecessors, is to publish probing histories and thorough editions of technical developments in mathematics and physics, broadly construed. Its scope covers all relevant work from pre-classical antiquity through the last century, ranging from Babylonian mathematics to the scientific correspondence of H. A. Lorentz. Books in this series will interest scholars in the history of mathematics and physics, mathematicians, physicists, engineers, and anyone who seeks to understand the historical underpinnings of the modern physical sciences.

More information about this series at http://www.springer.com/series/4142

Jan von Plato

Can Mathematics Be Proved Consistent?

Gödel's Shorthand Notes & Lectures on Incompleteness

Springer

Jan von Plato
Department of Philosophy
University of Helsinki
Helsinki, Finland

ISSN 2196-8810 ISSN 2196-8829 (electronic)
Sources and Studies in the History of Mathematics and Physical Sciences
ISBN 978-3-030-50878-4 ISBN 978-3-030-50876-0 (eBook)
https://doi.org/10.1007/978-3-030-50876-0

Mathematics Subject Classification (2010): 03F40, 01Axx

This Springer imprint is published by the registered company Springer Nature Switzerland AG
The registered company address is: Gewerbestrasse 11, 6330 Cham, Switzerland

Preface

This book contains all that is found in Gödel's preserved shorthand notebooks on his research that led to the famous incompleteness theorems of formal systems. The notes are followed by the original version of his article, before a dramatic change just a few days after it was handed in for publication, and six lectures and seminars in consequence of his celebrated result published in 1931. The notebooks and one of the lectures were written in German Gabelsberger shorthand that I have translated into English, usually from an intermediate transcription into German, but at places directly. I thank Tim Lethen for his help in the reading of many difficult shorthand passages, and Maria Hämeen-Anttila for her support, especially at the troublesome moment when I discovered Gödel's tricky change of his manuscript after it had been submitted for publication. Marcia Tucker of the Institute for Advanced Study was very helpful during my visit to the Firestone Library of Princeton University where the originals of Gödel's manuscripts are kept. Finally, I recollect with affection my mother's decision to challenge her little boy by enrolling him in the German elementary school of Helsinki, a choice without which I would not have learned to read Gödel's manuscripts.

Jan von Plato

Acknowledgment

The Kurt Gödel Papers on incompleteness that this book explores are kept at the Firestone Library of Princeton University. A finding aid with details about their contents is found at the end of the fifth volume of Gödel's *Collected Works.* The papers were divided by their cataloguer John Dawson into archival boxes and within boxes into folders. Folders can have a third division into documents, with a running document numbering system. The papers have been mainly accessed through a microfilm that is publicly available, but also directly in Princeton. References to specific pages of notebooks usually require the use of the reel and frame numbers of the microfilm and that is how the sources are mostly identified in this book. Passages from Gödel in the introductory Part I are identified *in loco*. The shorthand manuscript sources on which Gödel's 1931 article is based are described in detail in Part II, Section 2 of this book. The typewritten sources for his 1930 summary and the 1931 article are described in the following Section 3. The sources of the six lectures and seminars on incompleteness are described in the last Section 4 of Part II. These descriptions together with the frame and page numberings in Parts III–V allow the interested reader to identify the source texts with the precision of a notebook page.

The writing of this book has been financed by the European Research Council Advanced Grant GODELIANA (grant agreement No 787758).

Contents

Part I

Gödel's steps toward incompleteness

J. von Plato, *Can Mathematics Be Proved Consistent?*, Sources and Studies in the History of Mathematics and Physical Sciences, https://doi.org/10.1007/978-3-030-50876-0_1

1. THE COMPLETENESS PROBLEM

David Hilbert's list of 23 mathematical problems from the Paris international congress of mathematicians of 1900 had as the second problem the question of the consistency of analysis: to show that no contradiction follows from the axioms for real numbers. A slip of paper with an additional problem, to be placed last as a 24th one, was found some hundred years later, one that asked for criteria for the simplicity of proofs and in general, "to develop a theory of proof methods in mathematics." The development of a theory to this effect, what Hilbert called *proof theory*, started in Göttingen in 1917–18, when the First World War was coming to its end. Its main aim was to provide answers to such questions as consistency.

After the war, German mathematicians were excluded from the international congress of mathematicians, held every four years. The reappearance of Germans on the international scene took place in the international congress of Bologna in 1928, with Hilbert lecturing on "Problems in the foundations of mathematics." In his lecture, Hilbert surveyed the development of mathematics in the past few decades, then listed four main problems in its foundation. There was behind the list the most remarkable period of research into logic and foundations of mathematics seen so far. Hilbert had realised that Bertrand Russell's *Principia Mathematica* of 1910–13 offered the means for formalizing, not just mathematical axioms as in geometry, but even the logical steps in mathematical proofs: "One could see in the completion of this grandiose Russellian enterprise of *axiomatization of logic* the crowning of the task of axiomatization as a whole." (Hilbert 1918, p. 153). Ten years later, the logic of the connectives and quantifiers had been brought to perfection, presented in the book *Grundzüge der theoretischen Logik* (Hilbert and Ackermann 1928). The formalization of arithmetic had also been accomplished, with recursive definitions of the basic arithmetic operations and an axiom system for proofs in arithmetic. Hilbert believed at this time Wilhelm Ackermann and Johann von Neumann to have solved the problem of consistency for a strong system of arithmetic, but there remained some doubts about it.

The first and second problems in Hilbert's Bologna list of 1928 are about the extension of Ackermann's proof to higher areas of mathematics. The list has as the third problem, from Gödel's reading notes on Hilbert's article:[1]

[1] Part of document 050135, reel 36, frames 377 to 385 in the Gödel microfilms.

III Completeness of the axiom system of number theory

i.e., to be proved:

1. $\mathfrak{S}$ and $\overline{\mathfrak{S}}$ to be shown as not both free from contradiction.

2. When $\mathfrak{S}$ is free from contradiction, it is also provable.

After a digression on the role of the principle of induction, there follows:

IV Completeness of logic (would follow from the completeness of number theory) "Are all generally valid formulas provable?"

So far proved only for propositional logic and the logic of classes.

Gödel had studied the Hilbert-Ackermann book in late 1928, then began to formalize proofs in higher-order arithmetic, for which purpose he invented an impeccable system of linear natural deduction. A whole long notebook, the "Übungsheft Logik" (exercise notebook logic) is devoted to this purpose, with formal derivations of unprecedented complexity, more than eighty steps and up to four nested temporary hypotheses made (cf. von Plato 2018b).

Gödel's interest shifted soon from the actual construction of formal derivations to the completeness of a system of proof. The completeness of quantificational logic is clearly formulated, independently of arithmetic, in Hilbert-Ackermann. There is a shorthand notebook with the title "Diss. unrein" (Dissertation draft), fifty pages long, with an outline of the first ten chapter headings on page 16, slightly abbreviated here (document 040001):

1. Introduction
2. Notation and terminology
3. Basic theorems about the axiom system
4. Reduction to denumerable domains [Denkbereiche]

...

7. Independence of propositions and rules
8. Extension for the case in which = is incorporated

9. Extension for axiom systems with finitely and infinitely many propositions

10. Systems with a finite basis and monomorphic systems

These items from 1 to 10 are detailed in the rest of the notebook.

Gödel's proof of completeness for the "narrower functional calculus," i.e., first-order classical predicate logic, has disjunction, negation, and universal quantification as the basic notions. The simplest case of quantification is the formula $\forall x F(x)$ with $F(x)$ a propositional formula. Gödel states in a shorthand passage that if such a formula is "correct," i.e., becomes true under any choice of domain of individuals and relations for the relation symbols of the formula, then the instance *with a free variable* x must be a "tautology" of propositional logic. In the usual "Tarski semantics" that is—unfortunately—included in almost every first course in logic, the truth of universals is explained instead by the condition that *every instance* be true, an explanation that with an infinite domain of objects becomes infinitely long.

In Gödel, by contrast, with the free-variable formula $F(x)$ a tautology, it must be provable in propositional logic by the completeness of the latter, a result from Paul Bernays' *Habilitationsschrift* of 1918 and known to Gödel from Hilbert-Ackermann. That book is also the place in which the rules of inference for the quantifiers appear for the first time in an impeccable form (p. 54, with the acknowledgment that the axiom system for the quantifiers "was given by P. Bernays"). With the free-variable formula $F(x)$ provable in propositional logic, the rule of universal generalization gives at once that even $\forall x F(x)$ is derivable. The step is rather well hidden in Gödel's completeness proof in the thesis that proceeds in terms of satisfiability. At one point, he moves to provability of a free-variable formula, then universally quantified "by 3," the number given for the rule of generalization.

Gödel's profound understanding of predicate logic, especially the need for rules of inference for the quantifiers without which no proof of completeness is possible, is evident through a comparison: Rudolf Carnap, whose course he had followed in Vienna in 1928, published in 1929 a short presentation of Russell's *Principia*, the *Abriss der Logistik*, but one searches in vain for the quantifier rules in this booklet. Other contemporaries who failed in this respect include Ludwig Wittgenstein and Alfred Tarski. The former was a dilettante in logic who thought that truth-tables would do even for predicate logic. With the latter, no trace of the idea of the provability of universals through an arbitrary instance is found in his famous tract on the concept of truth of 1935.

Gödel's actual aim in his doctoral thesis was a proof of completeness of arithmetic, as is witnessed by the last third of the planned contents of his dissertation. It should be noted that Hilbert's Bologna address that listed the problem got published after Gödel had finished the thesis:

> 11. Application to geometry and arithmetic – connection between the two – inclusion of functions over objects
>
> a.) for the case of completeness b.) for the case that no finite basis is at hand
>
> 12. General construction of resultants and solution of the problem whether real roots are at hand
>
> holds for which number systems?, decision procedure
>
> 13. Resolution of the Archimedean axiom, proof of the completeness of the arithmetic axiom system
>
> 14. There is no finite basis for arithmetic propositions
>
> 15. Independence of the concepts ?

There seem to be no traces of how Gödel in 1929 thought he would prove the completeness of arithmetic, though I have not studied the long notebook in every detail yet—perhaps the above already indicates some doubt? There is instead his announcement of the failure of any such proof the next year, found at the end of the lecture he gave at the famous Königsberg conference on the foundations of mathematics on 5–7 September 1930. The conference is remembered for its presentation of the main approaches to the foundations of mathematics, logicism, formalism, intuitionism, in three widely read lectures by Rudolf Carnap, Johann von Neumann, and Arend Heyting, respectively.

Gödel's short and readable lecture about the completeness of predicate logic—just twenty minutes were allotted for it— is preserved in shorthand and very slightly changed in a typewritten form that got first published in the third volume of Gödel's *Collected Works*. Close to the end of that lecture, we find the following passage (p. 28):

> If one could prove the completeness theorem even for the higher parts of logic (the extended functional calculus), it could be shown quite generally that from categoricity, definiteness with respect to decision follows.[2] One knows for example that Peano's axiom system is categorical, so that the solvability of each problem in arithmetic and analysis expressible in the *Principia Mathematica* would follow. Such an extension of the completeness theorem as I have recently proved is, instead, impossible, i.e., there are mathematical problems that can be expressed in the *Principia Mathematica* but which cannot be solved by the logical means of the *Principia Mathematica*.

It is clear from these remarks that Gödel's first thought was to extend the completeness result to higher-order logic, a point emphasised in Goldfarb (2005). The above is an indication of his way to the first incompleteness theorem from the time when the actual work was done, namely through a failed attempt that led to the insight about undecidability.

The shorthand version of the Königsberg talk ends with (reel 24, frame 311):

> I have succeeded, instead [of extending the completeness theorem to higher-order logic], in showing that such a proof of completeness for the extended functional calculus is impossible or in other words, that there are arithmetic problems that cannot be solved by the logical means of the PM even if they can be expressed in this system. These things are, though, still too little worked through to go into more closely here.[3]

In the typewritten version, we read somewhat differently about his proof of the failure of completeness (document 040009, page 10):

> In this [proof], the reducibility axiom, infinity axiom (in the formulation: there are exactly denumerable individuals), and even the axiom of choice are allowed as axioms. One can express the matter also as: The axiom system of Peano with the logic of the

[2] Literal translation of the German "Entscheidungsdefinitheit."

[3] The last sentence reads in German: "Doch sind diese Dinge noch zu wenig durchgearbeitet, um hier näher darauf einzugehen."

PM as a superstructure is not definite with respect to decision. I cannot, though, go into these things here more closely.[4]

Then this last sentence is cancelled and the following written: "It would, though, take us too far to go more closely into these things."[5] It would seem that matters concerning the incompleteness proof had cleared in Gödel's mind between the writing of the shorthand text for the lecture and the typewritten version.

The shorthand text for the Königsberg lecture is found fairly early in Gödel's two notebooks about incompleteness. There is, about sixty pages later, a shorthand draft for his two-page note on the two incompleteness results that he had prepared just before departing for Königsberg, with publication in October 1930. Whatever he had done about incompleteness by that point must have been before early September 1930, and some of it clearly earlier: Just a few pages before the Königsberg lecture text, Gödel writes that the formally undecidable sentences have "the character of Goldbach or Fermat," i.e., of universal propositions such that each of their instances is decidable. These examples suggest that a formally undecidable proposition $\forall x F(x)$ can have each of its numerical instances $F(n)$ provable, but still, addition of the negation $\neg\forall x F(x)$ does not lead to an inconsistency. Were the free-variable instance $F(x)$ provable, universal generalization would at once give a contradiction.

2. From Skolem's paradox to the Königsberg conference

Later in his life, Gödel gave various explanations of how he found the incompleteness results. He often repeated that he was thinking of self-referential statements, as in the liar paradox: *This sentence is false.* Replacing unprovable for false, one gets a statement that expresses its own unprovability. The explanation is good as far as it goes, and indeed given as a heuristic argument in Gödel's 1931 paper, but it gives little clue as to how one would start thinking along such lines in the first place. Gödel's meticulously kept notes and other material point at interesting circumstances that concern his discovery of the undecidable sentences.

As a first source from the time Gödel had begun work on incompleteness in the early summer of 1930 (by Wang 1996, p. 82; I would say perhaps

[4] The last sentence is: "Auf diese Dinge kann ich aber hier nicht näher eingehen."
[5] "Doch würde es zu weit führen, auf diese Dinge näher einzugehen."

May) there is Fraenkel's *Einleitung in die Mengenlehre* that, as is seen from Gödel's preserved library request cards, he had taken out in early April. Fraenkel discusses the question of decidability in principle of any mathematical problem, remarking that not a long time ago, every mathematician would believe in such solvability (p. 234):

> It is a fact that until today, no mathematical problem has been proved to be "unsolvable." The discovery of such a problem would without doubt present an enormous *novum* for mathematics, and not only for it.

Fraenkel is very clear about *Skolem's paradox:* The propositions of a truly formal system form a denumerably infinite class, and therefore in particular the provable propositions, i.e., the theorems. The seemingly paradoxical consequence is that formal (first-order) theories of real numbers and of set theory admit of interpretations in which the domain is only denumerably infinite. In particular, it can be taken to be the domain of natural numbers.

Further down, Fraenkel notes that "there should be nothing absurd in imagining that the unsolvability of a problem could even be *proved*" (p. 235).

A second early source bears the date 13 May when Gödel borrowed Skolem's "Über einige Grundlagenfragen der Mathematik." This 49-page article was published as a separate issue of an obscure Norwegian journal. There Skolem gives a striking version of his paradox: The denumerable infinity of propositions of a truly formal system can be *ordered lexicographically*. "Propositions about natural numbers," in particular, can be likewise thus ordered, but by contrast the properties of natural numbers cannot be so ordered, by which (p. 269):

> It would be an interesting task to show that every collection of propositions about the natural numbers, formulated in first-order logic, continues to hold when one makes certain changes in the meaning of "numbers."

Among the wealth of ideas in Skolem's paper, there is an outline for a proof that the consistency of classical arithmetic reduces to that of intuitionistic arithmetic (p. 260), a result Gödel proved in 1932 through his well-known double negation translation.

Next to the library slips, two early notebooks give indications of Gödel's reading through his summaries of papers by others. It has turned out recently that these were begun around August 1931, when Gödel accepted the task of writing together with Heyting a short book on "Mathematical Foundational Research." The first notebook has two articles listed on each page, on top and half-way down, at times with notes, at times not, altogether over a hundred articles that relate to the topics Gödel was supposed to write as his part of the book project. Then there is the earliest preserved and clearly written-out notebook with the text *Altes Excerptenheft I* (1931—...) on the cover and a continuous page numbering (document 030079). This *Heft* gathers together some of his most important sources at a time when there were no photocopiers.

In his three-page summary of Skolem's long article (*Excerptenheft*, pp. 25–27), Gödel begins with Skolem's § 2, "proof of set-theoretical relativism" in Gödel's words, and then comes § 1, "enumeration of possible properties (therefore also sets) in Fraenkel's as well as Skolem's separation axiom." The last item in Gödel is for Skolem's §7, with the condition $ah - bk = 1$ pointing at the unique decomposition into prime elements in principal ideal domains:

> §7 Example of a domain that is not isomorphic with the number sequence even if it is an integral domain and even if for every two relatively prime h, k, $ah - bk = 1$.
>
> Conjecture that the number sequence is not at all characterisable by propositions of first-order logic.

At the end of this section, Gödel paraphrases Skolem's conclusion: "There is no possibility to introduce things nondenumerable as anything else but a pure dogma."

Gödel's summary was written down after his work on incompleteness had been finished and published. Still, Skolem's paper contains important ideas he had seen before that work. The way from these ideas to a first intimation of incompleteness is not long. One would likely think along the following lines:

Properties of natural numbers can be given as arithmetic propositions $F(x)$ with one free variable, and they can be listed in a lexicographical order, $F_1(x), F_2(x), \dots F_n(x) \dots$. Each of these properties $F_i(x)$ corresponds to a set of natural numbers, those for which the property holds and usually written as $M_i = \{x \varepsilon N | F_i(x)\}$. These sets form a denumerable sequence, but the sets of natural numbers as a whole form a continuum; each of them corresponds to a real number. The M_i give just a denumerable sequence of real numbers that one can diagonalise by the familiar argument of Cantor. Then we have a set D of natural numbers that is different from all of the M_i. Could we describe the diagonalization procedure within arithmetic itself, to form an expression in the language of arithmetic that corresponds to the diagonal set D, i.e., some free-variable formula $G(x)$ such that $D = \{x \varepsilon N | G(x)\}$?

To realise a possibility is one thing. To express provability in a formal system inside the system itself and to construct a proposition that expresses its own unprovability is, then, the real discovery. The Gödel notes show stages of the development of his ideas. The clearest turning point is one connected to the Königsberg conference. Before that, Gödel's argument was to give a truth definition for propositions of *Principia Mathematica*, then to prove that all theorems are true. If the proposition that expresses its own unprovability were provable, it would be true, hence unprovable, so it cannot be a theorem.

Gödel saw very clearly that the truth definition is the element in his proof that cannot be expressed within the formal system. He asked what it was that made his proof of undecidability possible. It was the said metatheorem about the truth of all the theorems, by which it could be decided that the constructed self-referential proposition is not simply false. If that decision could be made within the system, the unprovable proposition would follow. Therefore the truth of theorems is unprovable in the system.

The above argument is, in brief, a proof that the consistency of the system of *Principia Mathematica* cannot be proved within the system, or Gödel's original second incompleteness theorem. His later recollections dated its discovery to the times of the Königsberg conference. At that time, he prepared the mentioned short note of his results that appeared in October 1930, the

> *Some metamathematical results on definiteness with respect to decision and on freedom from contradiction*

This note was published in the *Anzeiger der Akademie der Wissenschaften zu Wien*, communicated by "corresponding member H. Hahn," Gödel's professor.

No trace of Gödel's original proof of the incompleteness theorems that uses a truth definition is left in his published article, but the idea surfaced from other quarters. Andrzej Mostowski knew Gödel from the late 1930s, from his stay in Vienna as recorded in Gödel's shorthand notes on the discussions they had. After the war, Mostowski became the author of the first book on Gödel's incompleteness theorems, the *Sentences Undecidable in Formalized Arithmetic: An Exposition of the Theory of Kurt Gödel* of 1952. There he describes two main ways of proving incompleteness, the first called *syntactic* and followed in Gödel's paper, the second *semantic*. The latter gives (p. 10) "an exact definition of what may be called the class of true sentences," with Gödel's theorem following from three conditions: "Every theorem of (S) is true," secondly the condition that no negation of a theorem be true, and as third the condition by which the truth predicate is equivalent to a condition of unprovability. A footnote on the next page states that "the idea of the semantical proof of the incompleteness theorem is due to A. Tarski," the long work on the concept of truth in formalized languages of 1935.

The second series of Gödel's notes contains, about six pages before the Königsberg break, the following (page 300R below):

> We go now into the exact definition of a concept "true proposition." The idea of such a definition has been expressed [cancelled: simultaneously] independently of me by Mr A. Tarski from Warsaw.

On the next page, we read:

> Now one arrives also quite exactly at proving (through complete induction) that
>
> *Each provable proposition is true.*

Tarski had visited Vienna in February 1930 and gave some lectures there that Gödel followed. A hint on their discussions is given by a letter Gödel wrote to Bernays on 2 April 1931. One finds there a "class sign" $W(x)$ read as "x is a *true proposition*," with truth of negation, disjunction, and universal quantification defined in the standard way (*Collected Works IV*, p. 96):

> The idea to define the concept of a "true proposition" along this way has been, incidentally, developed simultaneously and independently of me by Mr A. Tarski (as I gather from an oral communication).

The characteristic of Gödel's pre-Königsberg proof of the incompleteness theorems was that he used whatever means of classical mathematics, analysis and set theory included, in metamathematics. After the Königsberg meeting, the concept of truth and even the intuitive notion of "correctness" disappeared absolutely from his notes on incompleteness: one simply doesn't even find these words anymore, but instead an emphasis on the constructiveness of his proofs achieved through an "arithmetization of metamathematics" by elementary means. A hint of his original proof method is contained in the lectures Gödel gave on incompleteness in Princeton in the spring of 1934. There is a brief heuristic discussion of an arithmetic predicate $T(z_n)$ that expresses the "truth of the formula with number n," similar to the truth predicate W of his earlier writings.

There has been some lament about Gödel not acknowledging Tarski's approach to incompleteness. In the light of the above, the matter was *dejà vu* for Gödel, and not original to Tarski. From what has come out above, Gödel had begun work on incompleteness in May or June 1930. How does this fit together with Tarski's visit several months earlier, if the concept of "true proposition" was developed simultaneously? Gödel had arranged for an opportunity to discuss with Tarski and knew in that way about Tarski's ideas. In February 1930, he was in need of a truth definition for the aim that comes out so clearly from the Königsberg lecture, namely for the completeness of higher-order logic, the type theory of Russell's *Principia,* to be a well-posed problem. Such a concept would cover his system of proof in the 1928/29 *Übungsheft,* also to decide what axioms to accept in higher-order logic. The topic of a truth definition was of great systematic value for Gödel who mentions in his shorthand notes from the 1930s several times a folder named "The concept of truth" (Mappe "Wahrheitsbegriff").

3. From the Königsberg conference to von Neumann's letter

Among Gödel's audience in Königsberg sat Johann von Neumann who reacted at once and wanted more explanations. The two had discussions at the conference and in Berlin, where Gödel stayed for a few days immediately after the conference. The most detailed account of these events is Wang (1996), section "Some facts about Gödel in his own words," that describes the first approach to incompleteness as follows (pp. 82–84):

> I represented real numbers by predicates in number theory and

> found that I had to use the concept of truth to verify the axioms of analysis. By an enumeration of symbols, sentences, and proofs of the given system, I quickly discovered that the concept of arithmetic truth cannot be defined in arithmetic.
>
> …
>
> Note that this argument can be formalized to show the existence of undecidable propositions without giving any individual instances.

Gödel's words are different from those of his notebooks of 1930; The "verification of the axioms of analysis" means that a concept of truth is established by which the axioms turn out true and the rules of inference maintain truth. The formulation of 1930 was that each provable proposition of Russell's type theory is true.

Von Neumann suggested in the discussion to transform undecidability "into a proposition about integers." Gödel then found "the surprising result giving undecidable propositions about polynomials."

An edited account of the Königsberg discussion was published in the journal *Erkenntnis* (vol. 2, 1931, pp. 135–151). It contained also a brief summary of the incompleteness result by Gödel with the title "Nachtrag" (addendum, pp. 149–151), written some time in 1930/31. A typewritten version, not essentially different from the published one, is found in reel 24, frames 240–242.

Gödel's library loan cards show that he stayed in Berlin right after the Königsberg meeting and that he requested again Skolem's long paper of 1929, on 12 September from a library in Berlin. We are at the most crucial turning point in Gödel's work on incompleteness, the abandonment of the proof idea by which all theorems of the *Principia* are true, proved by methods of set theory. The first sign of this change is a set of 13 shorthand pages, 360L to 366L, in particular page 364L in which it is stated that the concept of "contentful correctness" can be restricted to instances of recursive predicates. These pages begin in exactly the same way as the final shorthand version and come close to the formulations in the introductory parts of the published article: "The development of mathematics in the direction of greater exactness has, as is well known, led to wide areas of it being formalized." Another sign of change from a set-theoretic approach that uses the concept of truth to one that uses primitive recursive arithmetic is that

Gödel writes ω-consistency instead of $\aleph_0$-consistency, the latter still found in the *Anzeiger* note handed by Gödel's account in on 17 September.

Von Neumann lectured from late October 1930 on in Berlin on "Hilbert's proof theory." Carl Hempel, later a very famous philosopher, recollected the excitement created, even evidenced by contemporary letters for which see Mancosu (1999). The account is (Hempel 2000, pp. 13–14):

> I took a course there with von Neumann which dealt with Hilbert's attempt to prove the consistency of classical mathematics by finitary means. I recall that in the middle of the course von Neumann came in one day and announced that he had just received a paper from... Kurt Gödel who showed that the objectives which Hilbert had in mind and on which I had heard Hilbert's course in Göttingen could not be achieved at all. Von Neumann, therefore, dropped the pursuit of this subject and devoted the rest of the course to the presentation of Gödel's results. The finding evoked an enormous excitement.

These are later recollections; It is known that von Neumann got the proofs of Gödel's paper around the tenth of January 1931. As we shall soon see, what von Neumann received during his lecture course are the October 1930 note with the first and second theorem stated, and the manuscript of section 4 of Gödel's paper.

One of the few known participants in von Neumann's lecture course was Jacques Herbrand. He was born in 1908 and received his education at the prestigious *Ecole normale superieure* of Paris. He finished his thesis *Recherches sur la théorie de la démonstration* at the precocious age of 21 in the spring of 1929. He went to stay for the academic year 1930–31 in Germany, first Berlin from October 1930 on, then Hamburg and Göttingen from late spring 1931 to July. These stays were in part prompted by his work on algebra, where Emil Artin in Hamburg and Emmy Noether in Göttingen were the leading figures. Herbrand's life ended in a mountaineering accident in July 1931.[6]

There is a letter of Herbrand's of 28 November 1930 to the director of the *Ecole normale* Ernest Vessiot in which he mentions von Neumann's "absolutely unexpected results," then writes that for now he will write about

[6] My *Formal Machinery Works*, section 8.3 on "two Berliners" contains a detailed account of Herbrand's stay in Germany and his relation to von Neumann.

the

> extremely curious results of a young Austrian mathematician who succeeded in constructing arithmetic functions *Pn* with the following properties: one calculates *Pa* for each number *a* and finds $Pa = 0$, but it is impossible to prove that *Pn* is always zero.

As noted above, the pre-Königsberg part of Gödel's second notebook mentions that undecidable problems can have "the character of Goldbach or Fermat." There is a difference, though, for Goldbach's conjecture, if false, can be refuted by a counterexample. With Gödel's undecidable propositions, it happens that each instance $F(n)$ of a property of natural numbers is provable, by which there is no counterexample. Still, $\forall x F(x)$, classically equivalent to $\neg\exists x \neg F(x)$, or the impossibility of a counterexample, need not be provable within the system. Gödel hardly thought that Goldbach's conjecture would be a "Gödel sentence."

Gödel states that he found the arithmetical form of incompleteness right after the Königsberg meeting. Here are his own words about the change (from Wang 1996, pp. 83–84):

> To von Neumann's question whether the proposition could be expressed in number theory I replied: of course they can be mapped into the integers but there would be new relations. He believed that it could be transformed into a proposition about integers. This suggested a simplification, but he contributed nothing to the proof, because the idea that it can be transformed into integers is trivial. I should, however, have mentioned the suggestion; otherwise too much credit would have gone to it.[7] If today, I would have mentioned it. The result that the proposition can be transformed into one about polynomials was very unexpected and done entirely by myself.

Herbrand's letter shows that von Neumann knew about the polynomial formulation–the "arithmetic functions *Pn*" for which $Pa = 0$ is provable for each number *a*–therefore the matter must have surfaced during their discussions in Berlin.

[7] The wording of Wang's notes seems somewhat awkward here, as if Gödel needed to protect himself against a priority claim by von Neumann, deceased two decades earlier.

Looking at the notebooks, one realizes that Gödel's "arithmetization of metamathematics" was initially that natural numbers can be used as the basic symbols of a formal system and that formulas then correspond to series of numbers. This representation appears first on page 294R:

> We replace the basic signs of the *Principia* (variables of different types and logical constants) in a one-to-one way by natural numbers, and the formulas through finite sequences of natural numbers (functions over segments of the number sequence of natural numbers).[8]

The famous Gödel numbering through the uniqueness of prime decomposition is seen first on page 293R, but just in the margin. There is no explanation of these expressions, $2^x 3^y 5^z 7^u 11^v$, $2^u 3^v$, and p_n, the last the n-th prime, by which they must be later additions.

Incidentally, Gödel's page 299L gives a clue to the origin of the idea of coding formulas and proofs through the uniqueness of prime decomposition: Gödel had used the numbers 0–7 as arithmetic representations of his basic signs, then needed an unlimited supply of numbers to represent variables of all finite types. He took numbers greater than 7 divisible by exactly one prime as propositional variables, and those divisible by exactly $k+2$ primes as variables of type k.

The cancelled page 329L, written well before the Königsberg meeting, develops the idea of Gödel numbering, with the comment that by the mapping of series of numbers to numbers through a product of powers of primes, "metamathematical concepts earlier defined that concern the system S, go over into properties and relations between natural numbers." This mapping is put aside, however, and series of numbers continue to represent formulas and proofs until the final shorthand version that was written after the Königsberg meeting. There, on pages 254L-R, Gödel writes that by taking products of powers of primes, "a natural number is associated in a one-to-one way, not just to each basic sign but also to each finite series of basic signs" – an idea described as "trivial" in Gödel's recollections about his meeting with von Neumann.

---

[8] The German is: Belegungen von Abschnitten der natürlichen Zahlenreihe mit natürlichen Zahlen. The English wording is from the printed article in Van Heijenoort (1967), as approved by Gödel.

Eight days before Herbrand's letter, von Neumann had written to Gödel about his proof:

> It can be expressed in a formal system that contains arithmetic, on the basis of your considerations, that the formula $1 = 2$ cannot be the endformula in a proof that starts from the axioms of this system—and in this formulation in fact a formula of the formal system mentioned. Let it be called $\mathfrak{W}$.
>
> ...
>
> I show now: $\mathfrak{W}$ is always unprovable in systems free from contradiction, i.e., a possible effective proof of $\mathfrak{W}$ could certainly be transformed into a contradiction.

Gödel must have explained how undecidable propositions are constructed to von Neumann in Berlin, not just a blunt statement of incompleteness, namely the way in which the provability of a formula in a system can be expressed as a formula of that system, and likewise with unprovability. In particular, the unprovability of a contradiction, say $1 = 2$, becomes expressed through an arithmetic formula.

Von Neumann writes next that if Gödel is interested, he would send the details once they are ready for print. He asks further when Gödel's treatise will appear and when he can have the proofs, with the wish to relate his work "in content and notation to yours, and even the wish for my part to publish sooner rather than later."

Herbrand had explained the post-Königsberg statement of incompleteness in terms of polynomials to Vessiot, and five days later he writes another letter, to his friend Claude Chevalley, in the worst handwriting imaginable, but full of sparkling ideas that seem to spring from nothing. In the letter, Herbrand explains von Neumann's presentation of the incompleteness theorem as follows:

> Let T be a theory that contains arithmetic. Let us enumerate all the demonstrations in T; let us enumerate all the propositions $Q\,x$; and let us construct a function $P\,x\,y\,z$ that is zero if and only if demonstration number x demonstrates $Q\,y$, Q being proposition number z.
>
> We find that $P\,x\,y\,z$ is an effective function that one can construct with arithmetic functions that are easily definable.

Let β be the number of the proposition $(x) \sim P\,x\,y\,y$ ($\sim$ means: not); let $A\,x$ be the proposition $\sim P\,x\,\beta\,\beta$

A the proposition $(x).A\,x$ ($A\,x$ is always true)

$A\,x$, equivalent to: demonstration x does not demonstrate the proposition β; so

$A\,x. \equiv .$ demonstration x does not demonstrate A

Let us enunciate:

$A\,x. \equiv . \sim D(x, A)$

1) $A\,x$ is true (for each cipher x); without it $D(x, A)$ would be true; therefore A; therefore $A\,x$; therefore $\sim D(x, A)$.

2) A cannot be demonstrated

for if one demonstrates A, $A\,x$ would be false; contradiction.

Therefore: $A\,0, A\,1, A\,2 \ldots$ are true

$(x)A\,x$ cannot be demonstrated <u>in T</u>

Next in Herbrand's letter comes the striking second incompleteness theorem. With $D(x,A)$ standing as above for "proof number x demonstrates proposition A," Herbrand writes in the letter the key formulas:

3) $\sim A \rightarrow D(x,A)$ et $D(z, \sim A)$

therefore: $\sim (D(x,A)$ et $D(z, \sim A)) \rightarrow A$

The conclusion, for the unprovable proposition A, is that "if one proves consistency, one proves A": Consistency requires that for any proposition A, there do not exist proofs of A and $\sim A$. This inexistence can be expressed as the formula $\sim \exists x \exists z(D(x,A)$ et $D(z, \sim A))$, or in a free-variable formulation, as $\sim (D(x,A)$ et $D(z, \sim A))$ for each x and z.

The contrapositive of Herbrand's formula 3) states that consistency implies A, a formulation taken over from Gödel as we shall see.

Let us now turn to Gödel's final shorthand version of the incompleteness paper. It occupies the first 39 pages of a notebook (document 040014), with a beginning that is very similar to the typewritten version. The impressive list of 45 recursive relations in the published paper matches a similar list of 43 items, some ten pages, followed by the upshot of the laborious work in the form of a theorem:

VI. Each recursive relation is arithmetic.

After the text proper of the manuscript for the article ends, there are two attempts at a formulation of a title, like this:

> On the existence of undecidable mathematical propositions in the system of *Principia Mathematica*
>
> On unsolvable mathematical problems in the system of *Principia Mathematica*

There follow five pages with formulas, recursive definitions of functions, elementary computations, and a stylish layout for a lecture on the completeness of predicate logic given in Vienna on 28 November. Next the title "Lieber Herr von Neumann" hits the eye, with an unfinished letter draft that contains:

> Dear Mr von Neumann
>
> Many thanks for your letter of [20 November]. Unfortunately I have to inform you that I have been in possession of the result you communicated since about three months. It is also found in the attached offprint of a communication to the Academy of Sciences. I had already finished the manuscript for this communication before my departure for Königsberg and had presented it to Carnap. I gave it over to Hahn for publication in the *Anzeiger* of the Academy on 17 September. [Cancelled: The reason why I didn't inform you in any way [written heavily over: didn't tell anything] of my second result in Königsberg is that the precise proof is not suited to oral communications and that an approximate indication could easily arouse doubts about [heavily cancelled: correctness] executability (as with the first) that would not appear convincing.] Concerning the publication of this matter, there will be given only a shorter sketch of the proof of impossibility of a proof of freedom from contradiction in the *Monatsheft* that will appear in January[9] (the main part of this treatise will be filled with the proof of existence of undecidable sentences). The detailed carrying through of the proof

[9] [Despite its name, the *Monatshefte* (monthly notices) appeared in four yearly issues. January has been changed into "early 1931."]

> will appear in a *Monatsheft* only in July or August. I can send you a copy [Abschrift] of – proofs of my next work in a few weeks.
>
> I shall include the part of my work that concerns the proof of freedom from contradiction in a manuscript, so that you can see from it to what extent your proof matches mine.
>
> The carrying through of the proof appears together with my proof of undecidability in the next volume of the *Monatshefte*. I didn't want to talk about it before a publication because this thing (even more than the proof of undecidability) must arouse doubt about its executability before it is laid out in an exact way.

There are eight pages between a first and a second version of the letter, filled with Gödel's attempts at formulating the second incompleteness theorem in various ways and how it should be proved, until another letter sketch that first repeats the remarks about the second theorem in the *Anzeiger* note and about the "carrying through of the proof in a near *Monatsheft*," then continues:

> Now to the matter itself. The basic idea of my proof can be described (quite roughly) like this. The sentence A that I have constructed and that is undecidable in the formal system S asserts its own unprovability and is therefore correct. If one analyses precisely how this undecidable sentence A could still be metamathematically decided, it appears that this became possible only under the condition of the freedom from contradiction of S. That is, it was strictly taken not A but $W \rightarrow A$ that was proved (W means the proposition: S is free of contradiction). The proof of $W \rightarrow A$ lets itself be carried through, though, within the system S, so that if even W were provable in S, then also A, which contradicts the undecidability of A.
>
> Concerning the meaning of this result, my opinion is that *only* the impossibility of a proof of freedom from contradiction for a system *within the system itself* is thereby proved. (I.e., one cannot pull oneself up by one's own bootstraps from the swamp of contradiction.) For the rest, I am fully convinced that there is [cancelled: a finite] an intuitionistically unobjectionable proof of

> freedom of contradiction for classical mathematics [added above: and set theory], and that therefore the Hilbertian point of view has in no way been refuted. Only one thing is clear, namely that this proof of freedom from contradiction has in any case to be far more complicated than one had assumed so far.

At the time Herbrand wrote to Vessiot, 28 November, the "absolutely unexpected results" he alludes to are perhaps an indication of von Neumann's version of the second theorem. By 29 November, von Neumann has read Gödel's letter of reply and that shows in Herbrand's letter to Chevalley of 3 December. Gödel had explained to von Neumann that the second theorem is proved by first showing an implication *within* the formal system. The details are found in the interim pages between the two letter drafts–with even references to the typewritten incompleteness manuscript. Here κ is any "recursive consistent class" of formulas:

> Let us now turn back to the undecidable proposition $17 Gen\, r$. We shall denote the proposition that "κ is free from contradiction" by $Wid(\kappa)$. For the proof of the theorem that $17 Gen\, r$ is unprovable, only the freedom from contradiction of κ was used (cf. 1.) on page 30). So we have
>
> $$Wid(\kappa) \rightarrow \overline{Bew_{\kappa}}(17 Gen\, r)$$

If now $Wid(\kappa)$ were provable within the system, also the unprovable proposition $\overline{Bew_{\kappa}}(17 Gen\, r)$ would, which is impossible. Gödel's first letter draft contains that he sends the part of his manuscript with these details to von Neumann.

In von Neumann's second letter to Gödel, of 29 November, he writes:

> I believe I can reproduce your sequence of thoughts on the basis of your communication and can therefore tell you that I used a somewhat different method. You prove $W \rightarrow A$, I show independently the unprovability of W, though with a different kind of inference that likewise copies the antinomies.

A letter of von Neumann's of 12 January 1931 sketches to some extent his proof, but how exactly his "somewhat different method" relates to Gödel's formulation of the unprovability of consistency is hard to say. Herbrand's second letter is clearly based on Gödel's formulation.

Von Neumann's lectures must have been widely attended, but I have been able to secure only Hempel, Herbrand, and B. H. Neumann, and very likely Gerhard Gentzen as participants. There is among Gödel's correspondence a postcard by Hempel, dated 15.IV.31, in which we read about the lectures:

> Perhaps Carnap has told you already that Mr von Neumann had used the last fourth of his lectures on proof theory in the past winter to present your research. He referred with great emphasis to the fundamental consequences your results have for the formalistic attempts at a proof of consistency by contentful methods weaker than what is contained in the system that is to be proved consistent.

In his second letter, von Neumann asks if Gödel has been able to decide "whether mathematics is incomplete or ω-inconsistent," then explains why he thinks that "each intuitionistic consideration can be formally copied." Next comes the conjecture by which "your result has resolved the foundational question in a negative sense: there is no rigorous justification for classical mathematics."

Gödel's correspondence contains a continuation of his exchange with von Neumann, a shorthand draft for a letter sent in January 1931 and filed under the letter M (miscellaneous M, multiple recipients):

> Dear Mr von Neumann
>
> Many thanks for your letter [of 29.XI]
>
> I send you over today the proofs of my article. I have limited myself in this part I mostly to the system of the *Principia Mathematica* and just indicated roughly the general result. I have even dedicated the main attention to questions about undecidability and presented the matter that concerns freedom from contradiction only to the point in which no doubt can remain about its executability. I mean, it is after all obvious to anyone who knows the formalism that all considerations employed in section 2 can be formalized in system P. I shall, nevertheless, naturally carry everything through in the second part.
>
> As concerns the question whether mathematics is ω-contradictory or incomplete, a decision should be possible in the present

state of things only in the direction in which one shows an ω-contradiction. There should likely not be any such and it would be as improbable for me as mathematics being contradictory. For a proof of ω-contradiction could, in all probability, be formalized and a contradiction derived from it. Apart from that, all intuitionists are likely convinced that mathematics is neither contradictory nor ω-contradictory.

As concerns the question of formalization of intuitionistic proofs, your considerations have not convinced me.[10] There is obviously to each intuitionistic proof a formal system in which it is representable, but there is (as I believe) no formal system in which *all* intuitionistic proofs are representable. The essential unclosedness [Unabgeschlossenheit] and extensibility [Erweiterungsfähigkeit] that is inherent to each formal system depends in the end on type theory.

The letter draft ends somewhat abruptly. As mentioned, there is a third letter of von Neumann's, dated 12 January 1931 after he had received the page proofs of Gödel's article together with, one can presume, the above letter. It contains an outline of a "somewhat shorter carrying out of the unprovability of freedom from contradiction."

4. The second theorem: "Only in a realm of ideas"

No second part of the incompleteness article ever appeared. When asked about the matter, Gödel would answer as in Van Heijenoort (1967, p. 616): "The prompt acceptance of his results was one of the reasons that made him change his plan." Paul Bernays comments on the incompleteness article that he received in page proofs, in a long letter to Gödel of 18 January, 1931. He says nothing about the last section 4 that presents the second incompleteness theorem, but writes instead that assuming, as von Neumann with whom he must have been in contact, that each finitary proof can be formalized, Gödel's incompleteness theorem VI already gives as a consequence the unprovability of consistency.

[10] [Several incomplete sentences follow, including that the formalization of intuitionistic proofs could be too complicated, with the comparison: precisely in the same sense in which, say, computational operations with too big numbers cannot be carried through anymore.]

In Gödel's first letter draft, he wants to assure von Neumann that he had both results, even mentioning Carnap as witness and quoting 17 September as the date he gave his October 1930 note to Hans Hahn who would communicate it to the Academy, and a promise to send a copy of the part of the manuscript that deals with the second theorem. Then come eight pages of attempts at a satisfactory formulation, and the second letter draft in which just the proofs of the incompleteness article are promised once they arrive.

The pages between Gödel's two letter drafts to von Neumann are his notes for section 4 of his incompleteness paper. An inspection of his typewritten manuscript shows that the last three lines of page 41 have been cancelled. They contain the beginning of his closing paragraph as in the shorthand manuscript. Pages 42–44, with typed page numbers in contrast to the rest of the manuscript, contain the added section 4. The first proofs have a Roman "I" added at the end of the title to indicate the first part of a two-part article, a paragraph that explains the second theorem added at the end of the introduction, and a long footnote on the second theorem added in another place. The original proofs have no mention at all of the second theorem before the added section 4.

The change in the manuscript has gone through the hands of Hahn, editor of the *Monatshefte*. That is shown by a copy in the Gödel correspondence of a note by Hahn that instructs the printing shop to "adjoin these pages to the manuscript of Dr Gödel," with a date that seems to be 24/11. A close look at Gödel's typewritten manuscript used by the printer shows that page 41 is a stencil copy, in contrast to the other pages, and that the cancellation of the three lines was done by Gödel who therefore must have included this page from his stencil copy of the entire manuscript with the new section. Von Neumann's letter was sent on Thursday, received on Friday or Saturday, after which Gödel had the new section finished over an obviously agitated weekend, for the following Monday the 24th. Von Neumann's second letter that discusses the details of that section, as contained in Gödel's letter drafts and the notebook pages between them, was dated the following Saturday.

A shadow is cast on Gödel's great achievement; There is no way of undoing the fact that Gödel together with Hahn played a well-planned trick to persuade von Neumann not to publish. In his letter of reply, he reproduced details from section 4, freshly written after von Neumann's letter, but he also included his short note of October 1930 that contains a statement of the second theorem. The latter would have been enough, but Gödel panicked at the prospect of von Neumann publishing his second theorem. The writing is quite nervous, with cancellations and additions all over. Moreover, the typewritten manuscript and the first proofs that reveal his trick must have caused him quite a stress; nothing he could send to von Neumann who would have wondered why the magnificent second incompleteness theorem is not even mentioned in the lengthy introduction, but appears only in a final section 4. He got at most that section in November and the page proofs for the entire article much later, around the tenth of January.

Concerning the October 1930 short notice to the Vienna academy, the last page of the shorthand manuscript instructs to add to page 1 a reference to this note. There is in the title of Gödel's article a footnote that points to it, without further explanations. The microfilms contain a typewritten copy with a stamp "Akademie der Wissenschaften in Wien, Zahl 721/1930 eingefangt: 21.X.1930." The wording of "Satz II" is well known:

> Even when one allows in metamathematics all the logical means of the *Principia Mathematica* (especially therefore the extended functional calculus with the axiom of reducibility or without ramified type theory and the axiom of choice), there is *no proof of freedom from contradiction* for the system S (and even less if one restricts the means of proof in some way). Therefore, a proof of freedom from contradiction of the system S can be carried through only by methods that lie *outside* the system S, and the case is analogous for other formal systems, say the Zermelo-Fraenkel axiom system for set theory.

With Gödel's *Anzeiger* note published, it is clear that von Neumann had no new result to publish, and there would have been no need for Gödel to change anything, at most mention the results in the short notice.

The formulation in the *Anzeiger der Akademie* also confirms what I said above, namely that Gödel's early metamathematics used strong methods. Moreover, the printed text mentions ω-consistency, but in the manuscript and in the notes before Königsberg, Gödel always wrote $\aleph_0$-consistency, the latter a distinctly set-theoretic notation.

The typewritten manuscript of the incompleteness article, with the typesetters' leaden fingerprints on it, contains three lines at the end of page

41, and the rest exists only as the last paragraph in Gödel's shorthand:

> To finish, let us point at the following interesting circumstance that concerns the undecidable sentence S put up in the above. By a remark made right in the beginning, S claims its own unprovability. Because S is undecidable, it is naturally also unprovable. Then, what S claims is correct. Therefore the sentence S that is undecidable in the system has been decided with the help of metamathematical considerations. An exact analysis of this state of affairs leads to interesting results that concern a proof of freedom from contradiction of the system P (and related systems) that will be treated in a continuation of this work soon to appear.

Gödel shows here a cautiousness the editor of his *Collected Works* Sol Feferman liked to emphasise about him, just "interesting results" about consistency. The thought of von Neumann publishing the second theorem must have haunted him and led to the hasty addition of a section on results so far "zu wenig durchgearbeitet" as he put his closing words in the Königsberg lecture. In fact, Gödel was unable to prove the second theorem to his satisfaction and no "Part II" of the incompleteness paper ever appeared, neither do the shorthand notes suggest any such work even in manuscript form.

With the above details in mind, let us take a skeptical look at Gödel's two-page section 4. The shorthand manuscript gives the second theorem and the standard beginning: "Proof." It is changed in the typewriting into: "The proof is in outline as follows," and in printing into "sketched in outline." The sketch is that for the proof of undecidability of formula $17Gen\,r$, "only the freedom from contradiction" was used. This observation is written on a separate numbered line as the impressive-looking formula

$$Wid(\kappa) \rightarrow \overline{Bew_{\kappa}}(17Gen\,r)$$

Three similarly displayed implications follow, as on pages 275R and 276L below, with the consequent changed successively by the way Bew_{κ} and $17Gen\,r$ were defined. Reminding that everything used in these steps is definable within the formal system, Gödel notes that the provability of $Wid(\kappa)$ within the system would at once lead to the provability of the unprovable proposition, which is impossible.

Gödel's lectures and seminars from the time after the great article was finished show still confidence in the second incompleteness theorem, such as in the long lecture titled "Über unentscheibare Sätze." The crucial point is to justify the informal argument by which the undecidable proposition *A* follows from consistency *W* (p. 875 of the lecture): "A proof of this fact can, as a more detailed investigation shows, be carried formally through from the axioms of the system $\mathfrak{Z}$." Here $\mathfrak{Z}$ is the system of classical first-order arithmetic of Hilbert and Bernays. The same comes out in more general terms in the popular talk Gödel gave in New York in April 1934. He writes there $C \rightarrow A$ for the statement that consistency implies the unprovable formula, then claims that "this proof can be formalized so we have $C \rightarrow A$ provable" (p. 18 of the lecture).

In a letter of 11 September 1932, Gödel corrects Carnap's suggested definition of the concept analytical and writes that "I shall give in part II of my work a definition of 'true' on the basis of this idea." In another letter to Carnap of 28 November the same year, he writes that "part II of my work exists only in a realm of ideas."

Part II

The saved sources on incompleteness

J. von Plato, *Can Mathematics Be Proved Consistent?*, Sources and Studies in the History of Mathematics and Physical Sciences, https://doi.org/10.1007/978-3-030-50876-0_2

1. Shorthand writing

Shorthand writing was regularly taught in high schools in Gödel's young years. The system was originally developed by Franz Gabelsberger in the 1830's; he called it "Redezeichenkunst," literally art of signs for speech. By Gödel's time, very many competing systems had sprouted from Gabelsberger, as well as adaptations to languages with phonetic patterns different from German. The idea was always to write down what was said, say at a law court, parliamentary session, or business negotiation. Thousands of lecture series from German-speaking universities have been saved as shorthand "Mitschriften" written down by students, or by the lecturers themselves and hidden in archives.

One of the principles of shorthand writing is that only that is indicated which makes a difference in spoken language. Therefore no regular punctuation marks are used, but Gödel uses often either a dash or longer spacing by which, in addition to sentence structure, it is seen that a new sentence and at times a new paragraph can be introduced. Often though, the fleeing nature of shorthand writing seems as if made for evading a firm grammatical structure of complete sentences. Systematic deviations from the phonetic principle include occasional quotation marks and underlinings, and the use of different sets of symbols that is a typical feature of any mathematical writing.

A stenographer would usually take notes, then produce soon after a polished typewritten text. One may wonder why anyone would use such writing for private purposes. The answer to this question, one I have had from many still active stenographic writers, is that once the habit of shorthand has been acquired, longhand writing is experienced as exasperatingly slow.

A word about the reading of shorthand sources is in place here: Their transcription has similarities to the recognition of spoken language it was originally planned to record; in both, understanding of words and phrase structure come hand in hand. With manuscript sources, there are in addition uncertainties for reasons such as faded sources, badly written or heavily cancelled passages, etc. At its worst, one can barely see a slight depressed line where there once was text, when the paper is illuminated obliquely.

Gödel's texts are full with cancellations, from one word to half a sentence to several pages in succession. The cancellations sometimes result in ungrammatical sentences that I have tried to repair by a reading of the unchanged and changed sentences, though if the cancellations are in an uncompleted sentence, the method does not work that well. Occasional comments on my part are found in square brackets, others in my footnotes. Gödel uses practically on every page parentheses and sometimes square brackets, with no clear difference. I have changed the latter into parentheses, except in formulas, to make clear who wrote what.

The source texts are on the whole often quite unpolished; I have tried to avoid the temptation to change that aspect during translation, in the direction of smooth English diction, trusting that Gödel's long and involved German sentence constructions can be turned into English with a reasonable level of readability maintained.

The translation of Gödel's German from around 1930 presents questions of terminology. One extreme is found in the accompanying material to his popular talk on incompleteness in New York, 1934, where words such as "beweisbar" were left in German by an unknown translator, for the lack of "precise English equivalents for the German words." Next "beweisbar" is explained as applying to a proposition that can be proved, as if there existed no word "provable" in English!

Gödel's New York talk, written by himself in English, gives some suggestions for translation. For example, he would usually write "free from contradiction," and I have translated correspondingly the German "widerspruchsfrei," instead of "consistent" that would be the standard usage today, except in "ω-widerspruchsfrei." A specific question comes with the words "Satz" and " Aussage." "Satz" can be translated as sentence or proposition, but it can also mean a theorem. Even "Aussage" is usually a proposition, as in "Aussagenkalkül," propositional calculus. Sometimes "Aussage" is best rendered as sentence, or even statement.

With formulas, it was usual in Gödel's times to write the letter symbols in what is known as Sütterlin-Schrift, a form of handwriting introduced in 1915. Gödel's notebooks from his early school years show how his writing changed from the earlier Kurrentschrift to Sütterlin. With foreign languages, at least in his Latin notebooks from the school and later with English, he would write in a way not different from any modern cursive handwriting. The Sütterlin alphabet is used for letters in formulas even in typewritten manuscripts, with symbols usually added in ink. A printer would set such symbols in fraktur. My translations follow these practices.

Words I could not read or guess with reasonable certainty are indicated by [?] and longer passages by [? ?]. The text can be very faint at places and especially the last lines of pages squeezed in a small space that is worn out and hard to read. Then there must be words I have read wrongly, but perhaps not too many, as the sentences I find usually make sense. There are still entire sentences the sense of which is not clear and would perhaps not have been even to Gödel himself. Such lacunae notwithstanding, one should still get a clear idea of how Gödel thought he would prove his incompleteness result, so I hope. As I wrote elsewhere (viz. in my 2018a, p. 4050):

> The transcription of shorthand is by the very nature of the script, with missing endings of words and abrupt shortenings—a single letter can stand for different words that have to be figured out from the context—also error-bound interpretation and guesswork.

The way from a transcription to an English translation is surprisingly robust; subtle differences in reading usually don't have any effect of note on a translation, and I judge the translations an adequate basis for a discussion of Gödel's views and their development. Scholarly work that concerns directly the unpublished writings requires, instead, the ability to read his Gabelsberger shorthand. Intermediate transcriptions into longhand German are not a fully adequate substitute for such purposes.

2. Description of the notebooks on incompleteness

1 § *The sources.* Gödel began around May 1930 the research that led to his famous article on the incompleteness of formal mathematical systems. Within half a year, or by mid-November, he had a finished typewritten 42-page manuscript that he sent for publication. His progress is recorded in two notebooks in which he wrote down versions of the article. The second notebook goes even beyond the date on which Gödel's manuscript was recorded as received by the journal *Monatshefte für Mathematik und Physik*, namely 17 November 1930.

The two preserved notebooks can be found among the Kurt Gödel Papers at Princeton University's Firestone Library, catalogued by Gödel's biographer John Dawson. The papers are divided into boxes and within boxes into numbered folders. There can be within a folder a third division by a running document enumeration, such as 040014 for the first incompleteness notebook.

The Gödel papers are available as microfilms that are usually quite readable, except for occasional faint pages or borders that have become too dark by the contrasty filming, with a last line sometimes hard to read for this reason. The incompleteness notebooks are found in reel 24, beginning with frame 245 with the explanatory text:

Folder 7
Undecidability Results (early drafts of *1931*):
AMs draft in 2 Notebooks, one
inserted in the other
"Unentsch. unrein"
written both directions,
filmed in original sequence
[1930?]

AM stands for autograph manuscript and TM is similarly used for typewritten manuscript. A document number 040014 has been stamped on three of the initial pages. The frames have usually a notebook opening with a left and right side, sometimes with what is indicated as an "intentional second exposure" to capture the text with a different intensity, especially if one side is an inside cover of a notebook with a darker background. Shorthand writing normally uses a pencil, with the consequence that some pages have become very faint with the years.

The first notebook has a blue cover and a box for text in which the title "Unentsch. unrein" (Undecidability draft) is written. The notebook has clearly recognisable unlined pages on which Gödel has written in a hand that is quite determined, with little cancellations, a sign of a final handwritten text for the article. (Other notebooks on squared paper of the school type have a less finished outlook.) The pages start with the cover on frame 248. The inside cover has additions to the text to follow, the latter starting with a heavily underlined "Arb. unentsch. unrein," a section number **1**, and the famous opening sentence about the development of mathematics in the direction of greater exactness. The original closing paragraph is found on page 268L, followed by various pages to be detailed below until page 280L.

At this point, the text hits the notes written in backwards direction:

From the back, the inside cover of the notebook has "Schmidt bedanken" (Schmidt to be thanked) on top and a list of names and addresses, frame 287L, clearly ones to whom offprints should be sent. The pages that follow have been filmed in the order in which they appear from the front. In the order of writing, there is a long list of formulas with the predicate *Bew* and substitutions that start from the last page 286. Page 285R begins with the end of a plan of contents:

c) Substitution and *Bew* definable

d) Proof of the theorem

e) Extension of both theorems

5. The conditions from 3 concern especially finite formal systems

a) Description of finite systems

Point c) has been changed from 3, d) from 4. The list of formulas from the first backward page continues for the rest of this page. The next page 284L begins an addition to the final shorthand version for the incompleteness article that I have placed where the corresponding text can be found in the typewritten manuscript.

I have found pages in completely different places in the microfilms that clearly have been ripped off from this notebook, to be detailed below.

The second notebook mentioned in the catalogue description is found between the back cover and the first backwards page of the first notebook. It consist of just six pages: a back, an inside back cover, a leaf that separates the cover from the notebook, and just two pages on a squared paper. The text is unrelated to the theme of incompleteness but what is left of this notebook has the same document number 040014 stamped on one page as the first one.

The next folder is listed as follows:

Folder 8
Undecidability Results (early drafts of *1931*):
Notebook Draft
"Unentsch. unrein"

written both directions
[1930?]

The document number is 040015, printed only on the inside leaf of a thick notebook that is preserved in one piece. The cover is a shiny black on which one cannot write. The first frame 292 is the inside cover and the unlined leaf, the back of it 293L, and the text written on squared paper begins on page 293R. The last frame is number 335, after which we find:

Folder 8
Undecidability Results (early drafts of *1931*):
Notebook Draft
"Unentsch. unrein"
Backward Direction
[1930?]

The text begins again on an inside cover, frame 337, with a few words that mention the "strange circumstance" about consistency proofs. The text proper of the next version of the incompleteness paper begins on page 339R, with additions on the adjacent leaf. The frames go on until number 367 where the text hits the forward direction, frame 335.

2 § *Division*. My presentation of the notebook contents is divided into eight parts, in an order determined by what seems to be the order of writing of the notes, with the first notebook's contents coming last. The order of writing of the remaining notes is not straightforward, and different ideas could be had about that. The abandonment of the notion of truth as a basis for the proof is the most conspicuous internal dividing line, seen quite late in the notebooks. There are at least a couple of external fixed points: At frame 306, there begins the text for Gödel's Königsberg lecture. At 346, there is the text for the *Anzeiger* note of 1930. By Gödel's letter to von Neumann, he had finished the note and even shown it to Carnap prior to his trip to Königsberg. On the other hand, in the same letter a few days after November 20, he wrote that he was "in possession of the result you communicated since about three months," which should be about mid-August. He had apparently written his shorthand text for the Königsberg lecture in good time, as there are over sixty shorthand pages between frames 306 and 346. I present the notes except the final one in the order they are found among Gödel's papers and leave it to later investigation to improve on that.

The second notebook has the shorthand version of Gödel's address at the Königsberg meeting in the middle, at frames 306–311. There follow many pages of formulas, and a new start for an article at frame 318. The text soon hits the reverse direction, with yet another series that has in the middle the draft for the *Anzeiger* note. At the end of this notebook, there is a four-page shorter summary and list of contents for an article, and a series of thirteen pages of notes that are in many respects similar to the heuristic introduction of the published paper.

The first notebook comes very close to the final typewritten manuscript, up to a distinct closing paragraph at the end of section 3. There follow two letter drafts to von Neumann and in between a draft of section 4 of the published paper that together form a separate part of my presentation. The notebook with these materials has been used also in the reverse direction for results Gödel found later and that are incorporated in the manuscript, judging from the typewritten version. These few pages deal with results on predicate logic.

I describe these series of notes under headings taken from suitable initial phrases of significative beginnings in them, with one unavoidable repetition:

1. "Undecidability draft. We lay as a basis the system of the *Principia*"
2. "There are unsolvable problems in the system of *Principia Mathematica*"
3. "The development of mathematics in the direction of greater exactness"
4. "The question whether each mathematical problem is solvable"
5. "A proof in broad outline will be sketched"
6. "We produce an undecidable proposition in the *Principia*"
7. "The development of mathematics in the direction of greater exactness"
8. "Let us turn to the undecidable proposition"

Many cancelled passages are just false starts. Some are longer, but at once rewritten in an improved form. I have usually left such parts out of the translation. Other cancelled passages can be helpful for the understanding of what is to follow or may contain interesting observations not found elsewhere and would in that case be included.

1. "Undecidability draft. We lay as a basis the system of the *Principia*"

The first series of shorthand notes for Gödel's incompleteness article of 1931 is found in the Gödel microfilm collection, reel 24, frames 292 to 297L. The unlined leaf between the front cover and the notebook proper, page 292R, contains a heavily written title "Unentsch. unrein" (Undecidability draft). The upper right corner has: "Welche Symbolik? Einsetzungsregel etc neue Ausführung vorläufige Fassung" (What symbolism? Rule of substitution etc new execution preliminary version). A text about seven lines long follows, faint as if it had been erased, but readable enough to see that the contents are similar to those on several other pages:

> Finally, there results by the above method the strange circumstance that the freedom from contradiction of system *S* cannot be proved by the logical means that are contained in this system itself. I.e., even if one allows for the proof of freedom from contradiction all the logical means of the *Principia*, it is impossible, and even the more so if one allows only a part.
>
> The above result lets itself be extended correspondingly also to other formal systems, say Zermelo Fraenkel's set theory.

The page continues by giving general conditions for incomplete systems and an explanation of $\aleph_0$-consistency. The reference to the second theorem clearly shows that these pages have been written afterwards. A similar thing is seen in many other notebooks.

The two cover pages are similar to pages 346L and 346R in the same notebook, with unprovability of consistency, Zermelo-Fraenkel, conditions for the theorem, and $\aleph_0$-consistency mentioned. These two pages are the shorthand version of Gödel's October 1930 note on his two theorems, written before his trip to Königsberg. Page 348R describes again the "strange situation" with metamathematical theorems. On page 365R, in a set of notes used directly in the writing of the final version, we have: "From this follows the strange result that one cannot carry through a proof of freedom from contradiction for the *PM* even with all the logical means contained in the *PM*." The final shorthand version of the incompleteness article had originally a closing paragraph that is similar, cited above. This paragraph was replaced by an added section in the published paper, in which the "strange result" is mentioned that Gödel seems to have found right before the Königsberg meeting.

The inside cover, page 292L of the notebook, has some seemingly unrelated formulas and then the text, written exceptionally in ink:

> There are further formulas of the narrower functional calculus in S for which neither universal validity nor the *existence of a counterexample is provable.*
>
> This result holds especially also for the system of classical mathematics as this has been put up by J. v. Neumann in Math. Zeitschr.

The remark about predicate calculus is found as "result 3" on frame 319, with the startling continuation:

> It follows from 3 especially that one cannot solve the decision problem of the narrower functional calculus even with the ways of inference of abstract set theory known today.

This statement presupposes the answer to be negative, and Gödel clearly thought that a proof to that effect could not be elementary. Turing's proof of the undecidability of predicate logic in 1936 must have been a great surprise to him in this respect.

The two front cover pages are followed by two unorganised pages of notes in the notebook proper. A systematic text begins, what seems to be the very first draft for an introductory part of Gödel's article, and goes on for six pages, 294R–297L, "Wir legen im folgenden Untersuchungen das System der Principia mit Reduzibilitätsaxiom (...) zu Grunde." (We lay as a basis for the investigations to follow the system of Principia with the axiom of reducibility.)

The contents of the translations I give of this first series can be summarised as follows:

1. Two front cover pages with a clear indication of the unprovability of consistency within formal systems, written afterwards and similar to the October 1930 *Anzeiger* note (292R–293L), and the two remarks on page 292L.

2. Two pages of unorganised notes (293R–294L)

3. Six pages of systematic introduction, beginning with: "We lay as a basis for the investigations to follow the system of *Principia*" (294R–297L)

The right page next to 297L is blank, and a new systematic text begins on the right of the next opening, page 298R:

2. "There are unsolvable problems in the system of *Principia Mathematica*"

"Im folgenden soll gezeigt werden, dass es im System der Principia Mathematica auch bei Hinzufügung des Abzählbarkeitsaxioms (...) unlösbare Probleme gibt." (It will be shown in what follows that there are in the system of Principia Mathematica, even under the addition of the denumerability axiom (...) unsolvable problems.) This second instalment can be considered a detailed execution of the proof sketched in the preceding one. The text goes on in a determined way for some fifteen pages until it is interrupted by a shorthand version of Gödel's talk on the completeness of predicate logic in Königsberg in early September 1930, very closely equal to the preserved typewritten version from which the publication in his *Collected Works III* stems (frames 306L to the upper part of 311R). The Königsberg lecture text begins very conspicuously with: "Meine Damen und Herren!" It ends with a few words about incompleteness, as cited in Section 1 of Part I above.

After the Königsberg break, there follow about 13 pages of formulas between frames 311 and 318. These pages contain concepts that are used later in the *Heft*.

The contents of the translations I give of this second series can be summarised as follows:

1. About fifteen pages of systematic text, interrupted by the Königsberg lecture and beginning with:"There are unsolvable problems in the system of *Principia Mathematica*" (298R–305R)

2. About 13 pages of formulas (311L–318L)

3. "The development of mathematics in the direction of greater exactness"

A new version of the incompleteness article begins at frame 318L, with a section number 1 and heading: "Entscheidungsdefinitheit" (Definiteness with respect to decidability). The opening phrase is in translation "The development of mathematics in the direction of greater exactness has, as is known, led in the end to great parts of mathematics being formalized (as intended, even the whole of mathematics)." Section 2, with no title, begins on page 323R and introduces a system of elementary number theory *Z* alongside that of *PM* and von Neumann's set theory. In all:

1. Section 1 with the above title, a rather long introduction (318R–323R).

2. Untitled section 2 that describes the formal system of number theory *Z* and how metamathematical concepts are represented in arithmetic (323R–331R).

3. Statement of the main result for *Z* and its extensions (332R–333L).

4. Discussion of formalism and the meaning of the result (334L–334R).

5. Unsystematic notes on recursion equations (335R).

4. "The question whether each mathematical problem is solvable"

The text under this heading seems to be Gödel's third draft for an incompleteness article, counting instalments 1 and 2 as belonging together, written late in the summer of 1930 and preserved in about 38 pages between frames 339 and 357 of reel 24 in the Gödel microfilm collection. The notebook contains after these pages others that are early versions of the initial pages of the shorthand version of the published incompleteness article.

There are three preliminary fragmentary pages 337L, 337R, and 338L (double exposure 339L) before a systematic presentation begins. These are the front and back covers and the inside cover of the notebook, written in backward direction. The inside cover page 339L has three additions to the adjacent page 339R with which the text begins; they are incorporated in the text.

The earlier parts of this set of notes have many cancellations, about six pages altogether out of 15 between 339R and 346R, at one place, more than three pages in succession.

A list of contents of the translations that follow is:

1. A general description in four pages of the work, and of the formal system for which incompleteness is to be proved (339R–341R).

2. "The formal system laid as a foundation" is described in detail on four pages, but after half a page, there follow more than three pages in succession that are cancelled but still clearly readable (342L–343R). These are needed for the reading of what follows and are included.

3. A sketch for what axioms to include and detailed presentation of these, together with the rules of inference and the notion of provability (344L–345R).

4. Shorthand version of the 1930 *Anzeiger* note (346L–346R).

5. "We have so far defined a sequence of concepts," a recapitulation and discussion of the nature of metamathematics (347R–348R).

6. "We begin now the systematic presentation of the proof," definitions numbered 1–27 followed by 20 unnumbered ones (349L–356R).

7. Page 355R gives the theorem by which "each provable formula is true."

8. The last four pages 356L to 357R contain definitions that lead to the arithmetic proposition that states its own unprovability. If provable, it is true, hence unprovable.

5. "A proof in broad outline will be sketched"

As noted above, the notebook that contains what seems to be the third draft for an incompleteness article has, after the last page 357R of that series, the four pages 358L to 359R that consist of a two-page summary, a list of contents for an article, and one page that contains various formulas. The summary begins:

> In what follows, a proof in broad outline will be sketched by which the Peano axioms, together with the logic of the *Principia Mathematica* (with natural numbers as individuals), do not form a system definite with respect to decision, not even when the axiom of choice is included.

A list of basic concepts is given on the formula page 359R that relates directly to the post-Königsberg part of an early draft, page 315R in particular that contains the definition $Clsz(x) \sim Form(x) \,\&\, (E1y)\, y\, Frv\, a\, x$, very similar to the more detailed $Klsz(n) \sim Form(n) \,\&\, (E!x)\{x\, Frv\, a\, n\ x \leqq Höh(n)\}$ on page 359R. The list of contents is:

1. Introduction (for easier expression of the theorem)

2. System of number theory (here also finite sets at hand)

3. Extension

4. One-to-one association between formulas and numbers. Metamathematical properties = properties of numbers
5. Concept true
6. Concept finitely definable
7. Theorem expressed stated in different forms
8. Carrying through of the proof by the lemma
9. Proof of the lemma

The topics of these section headings appear in various combinations and orders in the successive drafts for the incompleteness article.

6. "We produce an undecidable proposition in the *Principia*"

The last-mentioned four pages are followed by a new set of notes 13 pages in length, pages 360L to 366L, with an opening sentence as in the published paper, about "the development of mathematics in the direction of greater exactness." The same opening is found in three places in the notebooks, and there are similar repetitions of other favourite formulations of Gödel's. To make a difference here, I chose as a descriptive title the opening phrase of page 361R that begins a more detailed exposition of Gödel's argument.

These thirteen pages are clearly written to be a first, introductory section for a paper on incompleteness. They follow the new ideas of September 1930, with the notion of truth put aside. The introduction of the last handwritten version is directly based on it.

7. "The development of mathematics in the direction of greater exactness"

Gödel's final set of shorthand notes is, as mentioned, written on a more prominent-looking notebook in comparison to the earlier ones. The text is mostly almost verbatim as in the typewritten manuscript, to the extent that the latter usually helps in clarifying difficult passages in the shorthand. There are substantial changes at two points to be discussed below. Gödel made extensive additions to the first proofs, but practically no deletions. Many formulas are indicated as being numbered, with space left for that as in ( ). The lacunae and missing cross-references can be gathered from the final typewritten manuscript and published article.

The footnotes in this version of the incompleteness article are usually an integral, planned part of the article. They are basically numbered consecutively, to note 29 at about page 25 of 42. Some, though, are later additions, as the first one that is signalled by[2)] in the text and found squeezed at the bottom of the page that follows. After Gödel lost count, he used asterisks, little squares, and other favourite graphical signs of his to indicate which footnote belongs where. I give the footnotes in the succession in which they appear in the main text, irrespective of the actual numbers or other signs they may bear.

The two places in which the notebook differs from the typewritten version are

1. The typewritten version (TM) has between pages 264L and 264R a discussion of whether the proof of theorem VI is intuitionistic (TM pp. 30–30a).

2. There is a gap between pages 267L and 267R. The typewritten manuscript gives here the theorem IX by which there exist undecidable arithmetic problems expressible within the narrower functional calculus, a result Gödel found relatively late.

The missing pages for the latter are found in two places: The back of the notebook, written in reverse direction, has the pages in the order of writing: 284L, 283L, 283R, and 282L. The continuation from the last is found in a completely different place, in reel 20 that has materials from the year 1936 when Gödel had his long period of recovery from a nervous breakdown. He stayed from the late summer on for long periods in the Austrian Alps, in a place called Aflenz, where he wrote mainly remarks on the foundations of quantum mechanics. The pages in question contain a prominent "black hole," an inkspot, and have been ripped off the main notebook between the mentioned pages 267L and 267R. Page 20-495R continues page 282L and is continued by page 267R. Page 20-496 is the backside of page 20-495R and seems to be what at some point was meant to close the formal development. There is a loose page filmed as 20-495L that has notation and definitions that continue on the right side 495R. I have placed these pages in the order dictated by the typewritten manuscript.

The text proper of the notebook ends with a final paragraph on page 268L, as well as some corrections and additions to the typewritten manuscript, followed by two attempted titles mentioned above, and some recursion formulas on the next page.

Next in the notebook come two pages of attempts with formal notation, and then a page with a plan of contents for a lecture on completeness that Gödel gave in Vienna on 28 November 1930.

8. "Let us turn to the undecidable proposition"

After another two more pages of formulas, there comes in the notebook with the final shorthand version Gödel's first draft of a letter of reply to von Neumann's letter dated 20 November, pages 272L and 272R. There follow eight pages, 273L to 276R, with attempts at a proof of the second theorem, including as a part the text for section 4 of the published paper. The second, final letter draft to von Neumann takes four pages, 277L to 278R.

Page 279R has a very clearly written statement in which Gödel writes that his results "stand in no contradiction with the Hilbertian formalistic standpoint." After a page with a few formulas, the four reverse direction pages are encountered. Three more pages of very densely written formulas follow. The incompleteness notes end with a list of names and addresses, reproduced below as an indication of who Gödel thought could be his prospective readers.

Inside the back cover, the remains of another notebook are inserted, with three back cover pages and just one leaf on squared paper, with various notes on Abelian groups, the history of logic, set-theoretic formulas, matrices, and continued fractions, unrelated to incompleteness.

The list of names at the back of the final incompleteness notebook begins with: "Schmidt bedanken," Schmidt to be thanked. Erhard Schmidt was an influential professor of mathematics in Berlin who made publicity for Gödel's results in the fall of 1930. The other names come with cities and street addresses, the latter omitted here. They are, in order of appearance:

Hempel Berlin, Dr Kohlenberg [?] Berlin, Doz. W. Dubislaw Berlin, Reichenbach ? Berlin, Dr A. Heyting Enschede, Schlick, Doz. Behmann, Prof. P. Bernays Göttingen, Hilbert, Noether, Dr Ackermann Münster, Skolem, Tarski, Presb.[urger] Jerusalem, Scholz.

The physical appearance of the pages in my eigth insertion of the Gödel incompleteness notes hides behind itself the following order of writing:

1. First Gödel writes a draft for an answer to von Neumann, the two pages 272L and 272R.

2. Next he writes four pages on how the second theorem follows from his considerations in the manuscript handed in for publication on 17 November 1930, the pages 273L–274R.

3. Now there follows the text of Gödel's new section 4, practically *verbatim* the published version, on pages 275R to 276R, with the above-mentioned inserted remark about the relation of the unprovability of consistency to Hilbert's program on the isolated page 279R.

4. The second letter draft to von Neumann in four pages, 277L to 278R.

3 § *Summary overview of Gödel's shorthand manuscripts.* The above division shows four successive shorthand versions of the incompleteness article:

1. The first version consists of the introduction and execution in instalments 1 and 2.

2. The second version is given in instalment 3.

3. The third version is given in instalment 4. It is followed by a plan for a new version in instalment 5, still based on the concept of truth, but not carried through.

4. The fourth and final version is given in instalment 7. Instalment 6 is a preliminary version for its introduction.

3. The typewritten manuscripts

1 § *The Anzeiger note of October 1930.* Gödel had written, by his letter to von Neumann, a short summary of his two incompleteness results before his departure for Königsberg on 3 September 1930. By the same letter, he gave the note in for publication on 17 September. As mentioned above, there is a shorthand version of this note at frame 346. The typewritten manuscript of this two-page note in the *Anzeiger der Akademie der Wissenschaften zu Wien* has been kept in the Gödel papers (box 7a, folder 9, document 040016). A title page has a stamp by which it has been received on 21 October 1930, communicated by "corresponding member H. Hahn" and the title "Einige metamathematische Resultate über Entscheidungsdefinitheit und Widerspruchsfreiheit."

Gödel' summary has the handwritten additional sentences, the first in his, the second apparently in Hahn's hand:

Theorems I, III, IV, can be extended even to other formal systems, for example the Zermelo-Fränkel axiom system of set theory.

The proofs of these theorems will appear in the *Monatshefte für Mathematik und Physik*.

The publication, in issue 19 of the *Anzeiger*, pp. 214–215, tells that the work was presented at a session of the academy on October 23. In a first set of proofs, $\aleph_0$-consistency has been set by mistake as X_0-consistency. The final version has the notation ω-consistency. Gödel's note was soon published and he was able to send a copy to von Neumann along his letter of about 25 November 1930.

2 § *On formally undecidable propositions, earlier version.* Gödel handed in his manuscript on 17 November 1930. The Gödel papers contain the printer's version, with a first page on which we find the title *Über formal unentscheibare Sätze* (box 7b, folder 12, document 040020). The first word is in shorthand. The next page is for the typesetter and instructs to print underlined words with spaced letters (Sperrdruck) and those underlined in green in italics, as in the final article. I have presented these as small capitals and italics, as in the Van Heijenoort translation. The same short title with the indication "earlier version" is found at the back of the proofs.

The typewritten manuscript follows closely the final shorthand version, except for the added section 4 on the second incompleteness theorem. One remarkable addition is a discussion, about half a page, that follows the proof of the central theorem VI about the existence of undecidable propositions. The argument is that if the undecidable proposition 17 *Genr r* were provable, an effective proof of ω-inconsistency would follow. The results about predicate calculus in Gödel's theorems IX and X were late achievements, not yet in a final form in the shorthand version.

The numerous footnotes of the last shorthand version have been turned into sixty-odd footnotes that I was able to reproduce with their original numbering maintained.

Even if I made the transcription and translation of the shorthand version first, following the order of writing, I made some use of the typewritten manuscript in the transcription. To have identical passages translated identically, the basis for the translation of the typewritten manuscript was that of the shorthand version. The process of comparison of the original shorthand and its transcription and translation with the typewritten manuscript led, in turn, to some changes and improvements in the other direction, and may have left a passage or two in which the wordings are not preserved precisely, with no effect on the content.

The typewritten manuscript has some changes, usually small ones, that I have followed to arrive at the precise version Gödel sent in for publication. These changes are readily visible to anyone who looks at the originals. They were preceded by changes revealed through notes in an odd place: at reel 44, frames 870 and 871, amidst all sorts of loose notes, there pops up a list of changes to the manuscript, with clear page indications that match the extant typewritten manuscript. Gödel had left the pages of this manuscript rather short, so could add footnotes at the bottom, either typewritten or done by hand. All of these are rather brief, such as the reference to von Neumann's 1928 paper on set theory on the first published page, except for two substantial additions. The first such addition is to page 30 of the TM, effected by a rewriting of this page and an added shorter page numbered 30a. The text to be added is (to be compared to p. 30 of the TM):

> The proof of theorem VI was carried through without express consideration of intuitionistic requirements. It is, though, easy to convince oneself that the following is shown in an intuitionistically unobjectionable way: If a formal decision (from κ) is presented (for the proposition effectively presentable), then one can effectively specify:
>
> 1. A proof for *Neg Gen r*.
>
> 2. A proof for $Sb(r_n^{17})$ for each arbitrary n.
>
> I.e., a formal decision of *Gen r* would have as a consequence an ω-contradiction that can be effectively shown (and this for each arbitrary recursive class that is laid out).

The second addition is to page 38, footnote 55, in relation of theorem IX:

> This does *not* stand in contradiction with the result of my work "Über die Vollständigkeit der Axiome des logischen Funktionenkalküls," because there it was proved only that each formula is either provable or there exists a counterexample.

At the end of the introduction to his article, Gödel writes about the two conditions his result needs: first, that the concept of "provable formula" can be defined within a system, and, secondly, that "each provable formula be contentfully correct." The comment is that "the exact carrying through of the above proof has as its main task to avoid completely the contentful interpretation of the formulas of the system considered." This has been changed into: "has as its tasks among others the substitution of the second of the conditions presented by a purely formal and much weaker one." Another cancellation in the introduction is on the first page, as explained below.

The main text, after the introduction, has just two essential cancellations: The first is the second footnote on page 262R (note 42). The second one is a passage on page 264R, found below in the footnote on that page. The footnote numbering follows the manuscript.

The closing paragraph has been substituted by a final section 4, pages 42–44 of the manuscript.

Gödel made extensive additions to the published paper in the first set of proofs. The two references to literature on the first page have been greatly extended by additional references to von Neumann, Hilbert, and others in Göttingen. Footnote 9 is a long addition, and at the end of the introductory section there is a new paragraph about the second theorem. Yet another long addition is at the end of section 2, with a reference to "part II of this work." The changes in the page proofs are, instead, mainly corrections of misprints, so that the modifications in Gödel's article as sent in and as published can now be found out by a comparison.

The added section 4 has in the first proofs a long new paragraph, before the closing paragraph.

4. Lectures and seminars on incompleteness

Gödel is known to have presented his incompleteness result in Vienna on two occasions prior to its publication: first in the Schlick circle on 15 January 1931, and then one week later, 22 January, in Menger's mathematical colloquium (see Dawson 1997, p. 73). Discussion remarks on Gödel's presentation have been preserved, prepared by the circle's court reporter Rose Rand. The German text is found in Rand (2002). The typewritten three-page manuscript of Gödel's 1932 publication "Über Vollständigkeit und Widerspruchsfreiheit" (On completeness and freedom from contradiction) has at the back of the last page the text "Menger Kolloquium" followed by "report on my works" written in shorthand.

Gödel's papers contain a folder with a slip of paper, in part in shorthand, with the following (reel 24, frame 143):

> Manuscripts
>
> Proofs [cancelled: own works lectures]
>
> the 3 works in Mo. H. + Vienna lectures on the first two

The reference is to the *Monatshefte für Mathematik und Physik* in which Gödel published his completeness and incompleteness papers, and a third paper on the decision problem of predicate logic in 1933. There is a lecture on completeness placed with the slip (reel 24) but no lecture on incompleteness. The completeness lecture is ten pages long and with the title handwritten on the back of the last page: "Vortrag über Vollständigkeit des Funktionenkalküls" (Lecture on the completeness of the functional calculus). The text is virtually the same as the shorthand text for the lecture in Königsberg in one of the two notebooks on incompleteness. As mentioned, Gödel held on 28 November 1930 a lecture on completeness in Vienna. A list of contents is found in the final notebook for the incompleteness paper, after the text had been completed but before the letter draft to von Neumann. This list is quite different from what the preserved lecture contains. The idea is not far-fetched that Gödel had little time to prepare a new talk on completeness, his time before the talk being taken by the writing of a new section for the incompleteness paper and of an answer to von Neumann's letter of 20 November.

There are among Gödel's papers three carefully written German texts for lectures on incompleteness, similar in content but of varying length and detail, presented below in 1 § to 3 §. The first and second of them can be identified as a short and long version of the talk Gödel gave in a mathematics conference on 15 September 1931.

In 1931/32, Gödel was very active with Hahn's logic seminar and presented his completeness and incompleteness results there. The notes for these presentations are found in two folders, usually extensive but less polished. The cover of the first has: "1931/32 Hahn Logikseminar und Vortragsentwürfe" (logic seminar and drafts for talks). The second folder has the text "Hahn Logik-Sem 31/32 (Boschan)." The name refers to the mathematician Paul Boschan or von Boschan. The Boschan folder contains two very clearly written shorthand manuscripts, the first with the title "Vollständigkeit des Funktionenkalküls", the second with the title "Über unentscheidbare Sätze." The latter consists of three "Kanzleiformat" sheets, eleven big pages, and a separate sheet with the crucial formulas of the presentation. The presentation follows closely the Bad Elster typewritten versions, but with much more explanation and discussion.

There is in the Hahn folders also a typewritten manuscript that records the seminars, over 60 single-spaced pages with 22 seminar-hours dated between 22 October 1931 to 4 July 1932. I assume Boschan to be the one who prepared these. Gödel's papers contain two sets of these notes, with slight differences on some pages together with Gödel's handwritten small additions and changes. These are found in reel 25, frames 4 to 62. Most of the seminar notes deal with various aspect of propositional logic, then Gödel's completeness theorem, Herbrand's results, and finally on four last pages the incompleteness result, with the title "Über die Unmöglichkeit von Widerspruchsfreiheitsbeweisen" (On the impossibility of proofs of freedom from contradiction).

Gödel's first stay in Princeton was from the fall of 1933 to the beginning of summer 1934. He held lectures there on the incompleteness theorems, between February and May, that were published in a mimeographed form in 1934 under the title *On undecidable propositions of formal mathematical systems.*[11] In addition to these systematic lectures, he gave two general lectures on his theorems in April 1934, one in New York on the 18th, and another in Washington on the 20th. The titles were *The existence of undecidable propositions in any formal system containing arithmetic* and *Can mathematics be proved consistent?*, respectively. These two lectures are markedly different in emphasis. The former was meant for a general audience, whereas the latter, the manuscript of which was identified in the summer of 2019, clearly assumes some knowledge of mathematics, say the prime decomposition of natural numbers that appears in a detailed discussion of Gödel numbering and its use in the arithmetic representation of a Gödelian undecidable proposition. The same amount of detail on this crucial discovery is nowhere else to be found in Gödel's writings.

[11] The lectures became generally available through their publication in Davis (1965).

Gödel's original English writing of the two lectures shows some of his early idiosyncrasies such as writing "allways." Some such aspects may be just ordinary spelling errors, corrected here, but those that are systematic are left as they are. The writing on the whole is very clear. Additions and cancellations lead only in a few places to questions of what is intended. Gödel seems not to have been sure what English expression to use for formal derivability in the Washington lecture, what would be the German "formale Herleitbarkeit," and left in several places generous space for an insertion of the proper word. The lecture in New York was accompanied by a translation of the introduction of his paper in which even "beweisbar" was left untranslated by an unidentified translator, but in the lecture, Gödel himself uses the word provable.

Some aspects of Gödel's longhand writing stem most likely from the habits of a shorthand writer. For example, punctuation marks are left out more often than not. A capital letter can tell that a sentence begins, but not always. Similarly to shorthand, singular and plural can be left implicit, to be decided from the context upon reading.

1 § *Bad Elster lecture on undecidable propositions.* The first of two lecture texts (document 040406, reel 30, frames 33–38) has five typewritten pages with the crucial formulas in Gödel's argument numbered (1)–(4). A handwritten text at the back of the last page gives the title as: "Über unentscheidbare Sätze (Vortrag ?)" (On undecidable propositions (lecture ?)).

Gödel gives in the lecture a very clear description of the idea that led him to incompleteness, the diagonalization of one-place arithmetic predicates $\varphi_1(x), \varphi_2(x) \ldots$ through $\varphi_n(n)$ and the collection into a class of those n for which $\varphi_n(n)$ is not formally provable, $\sim Bew\ \varphi_n(n)$ in Gödel's notation. The last one is a one-place arithmetic predicate, so we have some k such that

$$\varphi_k(n) \equiv \sim Bew\ \varphi_n(n)$$

This is Gödel's formula (3). With k in place of n, we get

$$\varphi_k(k) \equiv \sim Bew\ \varphi_k(k)$$

This is Gödel's formula (4) in which a proposition states its own unprovability.

2 § *On formally undecidable propositions* (Bad Elster, longer version). It appears from the shorthand remarks at the back of the last page of this nine-

page typewritten manuscript (document 040405, reel 30, frames 22–31) that it is Gödel's text for the yearly meeting of German mathematicians at Bad Elster, 15 September 1931. It is clearly an extended version of the previous one, the note "earlier version" written with the title notwithstanding: The crucial formulas are numbered as (5)–(9), in continuation of the numbering of the short version.

The Bad Elster lecture was Gödel's first presentation of the incompleteness theorem for a larger mathematical audience. At this point, his paper had been available for about half a year, and major authorities in logic had fully endorsed what he had accomplished, including von Neumann and Bernays. Thus, there was no reason for the kind of reserve Gödel had expressed in his letter of late November 1930 to von Neumann, about the difficulty of presenting the topic in a convincing way. Gödel's lecture is striking in the way it cuts the formal details of incompleteness to a few formulas. The main presentation describes incompleteness in broad terms as a consequence of formalization in which one-place arithmetic properties can be only denumerably infinite, in contrast to classes of numbers. Gödel seems even confident that the proof of the second incompleteness theorem would follow in the way of the first one. At the end of the presentation, he mentions the "strange" consequence of incompleteness that there must exist results of elementary number theory provable only by the means of analytical number theory. A cancelled last phrase mentions that this could be a further topic of investigation for Gödel.

Gödel's talk is listed in the *Jahresbericht* of the German mathematical association (vol. 41, p. 85) with the title: "Über die Existenz unentscheidbarer arithmetischer Sätze in den formalen Systemen der Mathematik" (On the existence of undecidable arithmetic propositions in the formal systems of mathematics).

Among Gödel's audience in Bad Elster there was the set theorist Ernst Zermelo, a presence that led to a famous exchange between the two, both at the conference and afterwards in letters found in the fifth volume of the *Collected Works*. At the back of the last page of Gödel's manuscript, there is written in shorthand:

1.) Zermelo

2.) [Cancelled: As a clarification to your remark] end of my talk work I believe I have explained sufficiently see also what I mean by a formal system

4.) [4 written over 3, then cancelled: Clarification] Zermelo talk delivered

3′ Skolemism

The bottom of the same page has the title "Über formal unentscheidbare Sätze, frühere Fassung" (On formally undecidable propositions, earlier version).

There were altogether five talks at Bad Elster at the session starting at 4 p.m., so perhaps half an hour each. Some of the presentations are just listed by title in the *Jahresbericht*, others instead come with a short paper. The notes at the back of Gödel's longer version indicate that he lectured from it, but chose not to publish even the shorter version.

Zermelo's talk, right after Gödel's, is given as "Über Stufen der Quantifikation und die Logik des Unendlichen" (On levels of quantification and the logic of the infinite). The *Jahresbericht* contains a three-page account of Zermelo's talk, the central point of which is that the idea of representing mathematics as a "fixed finite system of signs" is a "finitistic prejudice." Further, this "Skolemism" leads to the existence of denumerable models for set theory, to which Zermelo comments that "one can prove everything from contradictory premisses." A paragraph is devoted to Gödel's result (p. 87), with a reference to the Vienna academy summary of October 1930. Zermelo shows some caution here, apparently in awareness of the attention that Gödel's theorem had received, with the conclusion:

> The real question, whether there exist in mathematics absolutely unsolvable propositions, absolutely unsolvable problems, is in no way touched by such relativistic considerations.

The exchange between Gödel and Zermelo is found, with ample commentary, in the fifth volume of the *Collected Works*. Gödel remained polite in the exchange, but a letter to Carnap of 11 September 1932 shows what he really thought: "Have you already read Zermelo's senseless criticism of my work in the last *Jahresbericht*?"

3 § *On undecidable propositions*. This text is the second of the two carefully written lectures in the Hahn folders, what the latter part refers to in the folder title "Hahn Logikseminar und Vortragsentwürfe" (. . . and drafts for

talks, the copula in shorthand). The lecture is written on large "Kanzleiformat" paper with the longhand title *Über unentscheidbare Sätze* (box 7b, folder 14, no separate document number but identifiable as in reel 24, frames 867–876), with formulas on a separate sheet (frame 866). It is seen from Gödel's wording that the text is written for a lecture, possibly with a later article version in mind. Dawson mentions 28 November 1931 as a date on which Gödel gave a talk at the Austrian Mathematical Society in Vienna (p. 78, but could this be the 1930 talk also on 28 November?).

4 § *On the impossibility of proofs of freedom from contradiction.* Gödel spoke about his incompleteness result, especially the unprovability of consistency, in Hahn's seminar on 4 July 1932. It is the last item in the typewritten seminar notes. There are matching pages in the Hahn folders consecutively numbered 1–8 (reel 24, frames 750–757) mostly written in longhand and clearly used by Gödel for the seminar talk. The typewriting uses notation available on a keyboard such as a minus-sign for negation where Gödel has $\sim$. Many formulas are much more readable in handwriting than in the stencil copy of the typewritten version. I have made some use of this aid in the translation of Gödel's seminar.

Gödel's handwritten notes are preceded by a four-page shorthand account, frames 746–749, that appears to be the basis for the longhand notes. I give it here to the point in which it becomes similar to the latter:

> We have now come to know a proof for the freedom from contradiction of a certain subsystem of arithmetic. It is the widest known proof of freedom from contradiction that we have at all. The reason is that a proof of freedom from contradiction is, in a sense to be made precise, impossible for certain comprehensive mathematical systems. To be able to formulate and prove a theorem that concerns it, I have to first treat another related concept, namely that of completeness of formal theories.

Gödel has in mind here the well-known proofs of consistency of arithmetic with free-variable induction, as given by Ackermann, von Neumann, and Herbrand. The last, in particular, was extensively treated in the Hahn seminar and later in Gödel's mathematical notebooks.

Gödel's presentation of his incompleteness theorem at Hahn's seminar is preceded by his succinct presentation of Herbrand's proof, in what must have been a very long session. I include it as a prelude to Gödel's presentation of his own result, and as a witness to his thorough preparation for all the things he confronted in his research program.

5 § *The existence of undecidable propositions in any formal system containing arithmetic.* The New York lecture text was found among the Gödel papers, identified by John Dawson, but no text of the Washington lecture ("A manuscript for the former, but not the latter, survives." *Logical Dilemmas*, p. 103). The manuscript for the New York lecture is preserved in an orderly fashion and clearly indicated as "Vortr. New York" (reel 25, frames 623–654). There are just some formulas to be added, and some cancellations and improvements. As mentioned, a typewritten English translation of the introduction to the 1931 incompleteness article of unknown translatorship was prepared, together with some words of explanation.

6 § *Can mathematics be proved consistent?* The lecture notes were found in June 2019 by Maria Hämeen-Anttila, during a search of the manuscripts for Gödel's 30 December 1933 lecture *The present situation in the foundations of mathematics*, the final version of which is published in the third volume of Gödel's *Collected Works*. The pages were among those indicated as manuscripts for the *Present situation* lecture but divided apparently randomly in two different places and with no title or other indication. After a shorthand summary and sketches of formula pages to accompany the lecture (reel 25, frames 387–390), there follows a page numbered 4 (frame 391), but it has a Πx quantifier notation of the Princeton lectures on incompleteness, whereas the Washington lecture uses (Ex) as a primitive, notation Σx in the Princeton lectures. There follows page 4 of the Washington lecture manuscript (frame 392), with the suggestive phrases: "I want to deal with two of these questions to-night. The first concerns the freedom from contradiction of mathematics." Pages 11 to 24 (frames 393–406) follow in succession. The next frame 407 has page number 3 and just five lines of text that do not fit in the lecture, frame 408 is page 7 of the lecture, and frame 409 page 10. Frame 410 has the page number 19·1 and a cancelled text that does not fit anywhere in the Washington lecture:

> So it remains only to be proved that $\sim A$ is not provable either. Now if $\sim A$ were provable then also $\sim C$ were provable, for from this impl. it follows immediately that $\sim A \rightarrow \sim C$. So in this case it would be provable that a system is contradictory. So if

we assume that our system is not c

This page belongs to the New York lecture but is there replaced by another one numbered 19.1.

Frames 411–414 have the titles Supplement I to Supplement IV. They contain additions to the main text. There follow the three clean formula pages, frames 415–417. As with some other notes for lectures, the formulas to be cited during the lecture are indicated by empty space or a line.

Frame 419 begins a second set of pages, with the title (in German): "Talk US, most probably 1934 or $\overline{33}$." Now follow the seven missing pages of the Washington lecture: 1–3 (frames 419–421), 5–6 (frames 422–423), and 8–9 (frames 424–425).

The above-mentioned pages put together make up the complete Washington lecture, with just the title missing. Gödel's shorthand plan of the lecture is:

> 1·) What is a formal system?
>
> a) The fact that mathematics reducible to a few axioms and rules of inference, relation between logic and mathematics
>
> b) Exact language needed for that. Basic symbols, which combinations of basic symbols are propositions, each such combination expresses a specific assertion (true or false) $\sim, \rightarrow, E$
>
> c) Axioms and rules of inference mix (examples), what is a proof? *Each mathematical proof* can be carried through within the system
>
> 2·) Setting-out of the problem[12]
>
> I.) Freedom from contradiction, a proof for it seems at first circular, but justified by rules of inference having different degrees of reliability
>
> II.) Completeness, or at least for subsystems
>
> 3·) Both questions to be answered negatively for given systems of mathematics, but not independent of the specific form (even

[12] [Added below: only now possible]

if this form gives all the proofs). Of what kind are the propositions for which one can show that they are undecidable? Another formulation: arithmetic cannot be formalized in a complete way.

Part III

The shorthand notebooks

J. von Plato, *Can Mathematics Be Proved Consistent?*, Sources and Studies in the History of Mathematics and Physical Sciences, https://doi.org/10.1007/978-3-030-50876-0_3

1. Undecidability draft. We lay as a basis the system of the *Principia*

292R[1]

What symbolism? Rule of substitution etc new execution
preliminary version

Undecidability
draft

Finally, there results by the above method the strange circumstance that the freedom from contradiction of system *S* cannot be proved by the logical means that are contained in this system itself. I.e., even if one allows for the proof of freedom from contradiction all the logical means of the *Principia*, it is impossible, and even the more so if one allows only a part.

The above result lets itself be extended correspondingly also to other formal systems, say Zermelo Fraenkel's set theory.

293L

This proof of undecidability can be widened to all extensions of system *S*, i.e., systems that arise from it through the adjunction of new axioms, as long as these fulfil the conditions:

1. The class of axioms added is finitely definable (this holds especially for all finite classes that are obtained from *S* through type elevation).

2. All provable propositions are true.

Condition 2 cannot be replaced by the condition of freedom from contradiction, i.e., there are extensions of the system of the *Principia* with finitely definable axiom classes in which not all propositions are definite with respect to decision. But such contain always false propositions.

[1] [This page and the next one are the inside cover leaflet pages, surely written after the main text that starts on page 294R. The acid paper has turned dark with the consequence that the writing is very faint, but one can see that the wordings are ones that appear elsewhere relatively late in the notebooks. The inside cover page 292L has the remarks:

> There are further in formulas of the narrower functional calculus in *S* for which neither universal validity nor the *existence of a counterexample is provable.*
>
> This result holds especially also for the system of classical mathematics as this has been put up by J. v. Neumann in Math. Zeitschr.]

One can instead replace conditions 1 and 2 by the following:

1. The class of axioms added is not only finitely definable but even definite with respect to decision, i.e., it is decidable for each formula whether or not it is an axiom.

2. The system is $\aleph_0$ consistent.

Here the meaning of $\aleph_0$ consistency is the following: for no property f of natural numbers that occurs in the system is simultaneously provable $f(1)\, f(2) \ldots f(n)$ *ad inf* and $(Ex)\overline{f(x)}$. There are extensions of the *Principia* where this fails.

293R

We replace the basic signs of the *Principia* (logical constants and variables of different types) in a one-to-one way by natural numbers, and correspondingly the propositions of the *Principia* through finite sequences of numbers. By "formulas" are, then, in the following always meant finite series of numbers (with certain precisely given properties). The purpose of this replacement is that by it, many propositions about the system of the *Principia* become expressed metamathematically within this system itself, because they are propositions about finite series of numbers.

Metamathematical concepts used in the following[2] $2^x 3^y 5^z 7^u 11^v$

[cancelled: formula], proposition $2^u 3^v$

[cancelled: class sign, true formula] p_n

Bew formula

class sign

[cancelled: proposition from a class sign]

[cancelled: ordering of the class signs]

finitely definable *negation*

Theorem: *Bew* is finitely definable, class sign, proposition from a class sign, ordering [of the class signs] as well,

[2] [The products of successive primes appear to be later additions close to the margin.]

294L

therefore also $\overline{Bew[F(n);n]} = K(n)$

i.e. $W[a;n] \sim K(n)$

a is a class sign, therefore $a = F(m)$

The proposition $[a;m]$ is undecidable

because from $Bew[a;m]$ follows

$W[a;m]$

$K(m)$

$\overline{Bew[F(m);m]}$ i.e. $\overline{Bew[a;m]}$

$Bew\,N[a;m]$

$W[N[a;m]]$

$\overline{W[a;m]}$

$\overline{K[m]}$

Bew

There is no difficulty in writing the undecidable proposition down *in extenso.* It has a relatively simple structure. It asserts (cf.) the inexistence of a proof figure for $[a;m]$. Here it is decidable for each proof figure, in a finite number of steps, whether it is a proof for $[a;m]$ or not.

294R [We lay as a basis the system of the *Principia*]

We lay as a basis for the investigations to follow the system of *Principia* with the reducibility axiom (but without ramified type theory), further the denumerability axiom (there are exactly denumerably many individuals), and with the axiom of choice (for all types). Instead of the denumerability axiom, one can postulate the Peano axioms for individuals (the relation of "successor" taken as an undefined basic concept).

We replace the basic signs of the *Principia* (variables of different types and logical constants) in a one-to-one way by natural numbers, and the formulas through finite sequences of natural numbers (functions over segments of the number sequence of natural numbers).[3]

[3] [What follows is cancelled: There occur in the *Principia* itself propositions about finite

We understand by "formula" a finite series of numbers with certain properties that can be given precisely and by "true formula" (Wa) one for which the associated proposition of the *Principia* is true. This concept is as unobjectionably amenable of mathematical definition as, say, a formula with 20 signs (cf. below p.).

A formula with one free variable of the type of the natural numbers in the *Principia* is a class sign. There exist naturally (just as formulas in general) only denumerably many and we designate by $F(n)$ the n-th relation sign in a determinate counting. We call[4] a k-place relation R between objects that occur in the *Principia* (i.e., classes and relations of a finite type) definable if there exists a formula with k free variables, of the kind that the proposition (formula) that arises through substitution of [cancelled: names of] objects

295L

is true if and only if the relation R obtains between these k objects. Finitely definable relations and properties are, briefly, those which can be defined within the system of the *Principia*. It is easy to convince oneself that the concepts formula, relation sign, further the counting relation, the relation F, as well as also the concept "provable formula" [written above: *Bew*] are altogether finitely definable.

We define now a [cancelled: relation R] class K of natural numbers in the following way:

2·)[5] $K(i) \underset{Df}{=} \overline{Bew(F(n);i\,0)}$ [$K(i)$ changed from $R(i\,k)$]

Since *Bew* and F as well as $(a;i\,k)$ cf. [?] are definable, then also $R(i\,k)$ [read: $K(i)$] i.e., there exists a formula a (relation sign) such that

1·) $W(a;i) \sim K(i)$ [changed from $W(a;i\,k) \sim K(i\,k)$]

a must occur in the counting F, i.e., $a = F(m)$ for a determinate m.

We claim now that the following proposition of the *Principia* is undeci-

---

sequences of natural numbers, and therefore the possibility is won to express a part of the metamathematical propositions in the system itself.]

[4] [A bullet at this place, in the margin, directs to the bottom of the preceding page, with a three-line footnote: If a is a relation sign and i, k natural numbers, we designate by $(a;i\,k)$ the proposition of the *Principia* (i.e., the finite sequence of numbers) that arises when one substitutes free variables in it by names of the numbers i, k.]

[5] [Number added in margin.]

dable:

$(a; m\,0)$ (there would be no difficulty in writing down this proposition *in extenso*).

295R

In fact, were $(a; m\,0)$ provable, then also true

$W(a; m\,0)$

But then there follows from 1·) in one direction [?]

$\overline{Bew(F(m); m\,0)}$ which states that

$(a; m\,0)$ is not provable.

So we have hit a contradiction. If we assume, then, that $\overline{Negation}(a; m\,0)$ is provable, then it would be also true, or we have $\overline{W(a; m\,0)}$. But from this would follow that $(a; m\,0)$ is provable, so $Neg(a; m\,0)$ not provable.

One recognises a close connection of this proof with the Richard antinomy, and it can be expected that even other epistemological antinomies can be reshaped into analogous proofs, something that in fact happens.

One can ask whether the system of the *Principia* could be so extended by the addition of new axioms that it becomes definite with respect to decision. As concerns this, one recognises that the above proof can be applied word for word even to each extension of the *Principia*,

296L

as soon as the following conditions are satisfied:

1·) Each proposition provable in it is true.

2·) The concept provable is finitely definable or in other words, the class of newly added axioms is finitely definable.

The latter turns up at any rate for each finite class of new axioms (also for each infinite one that arises through type elevation from a finite one) and one sees on the whole that a system that is definite with respect to decision and correct (no false propositions arise) could in any case be obtained only in an extraordinarily complicated way. The axiom rules would have to be so complicated that they would not let themselves be expressed within the system of the *Principia*.

What do the propositions $(a; m\,1)$ contentfully meant actually state? Obviously that there is no proof figure for a definite finite series of signs (= number sequence) that can be given. It is naturally decidable for each formula figure whether or not it is a proof figure for a series of signs in question. So $(a; m\,1)$ means the non-existence of a finite formula figure with a certain property that is definite with respect to decision.

296R

The theorem is, then, of the character of one like Goldbach's or Fermat's problem, so in a certain sense very simple, but nevertheless not decidable in the system of the *Principia*.

The exact definition of the class of "true formulas" anticipated above depends on the following: A proposition of the *Principia* is built up of elementary components of the form

> $\varphi(uv)$, $\psi(s)$ etc in which φ, u, v, ψ are variables of arbitrary types with only the restriction that $u\,v$ are suitable argument variables for φ and s a suitable variable for ψ

together with the operations ( ) $\vee$ $^{-}$. [Added remark: Formulas that contain propositional variables can be treated in a quite analogous way.] It lies therefore close at hand to define the concept true formula through recursion:

> 1·) If φ denotes the name of a class and u that of a suitable argument, then the

297L

> sign $\varphi\,u$ shall be called true when and only when u belongs to the class φ.
>
> $A \vee B$ shall be true when and only when either A or B is true.
>
> $\overline{A}$ shall be true when and only when A is not true.
>
> $(x)F(x)$ shall be called true when and only when $F(a)$ is true for each suitable name a.

This definition fails in that it assumes there to be names for all classes and relations which surely is not the case (for there are only denumerably many names). However, this can be easily remedied. For we consider the formulas not as spatial images but as abstract sequences of natural numbers. Nothing prevents us, therefore, instead of delivering the names of the classes and relations, to take the classes themselves. One must then consider also certain sequences of sets, sets of sets, etc of natural numbers as formulas. One sees easily how the above definition can be reformulated in this sense.

The theorem by which the class of provable formulas and the counting relation F are finitely definable becomes now a theorem that is quite exact and provable by usual mathematical methods. The same with the theorem that each provable formula is true.

297R [blank]

2. There are unsolvable problems in the *Principia Mathematica*

298R

There are unsolvable problems in the system of *Principia Mathematica*, as will be shown in what follows, and even under the addition of the denumerability axiom (there are exactly denumerable individuals) and the axiom of choice (for all types). Instead of the denumerability axiom, one can also postulate the Peano axioms for individuals, which results in the same and will be taken in what follows.[1]

As concern the logical auxiliary means in the following proof, no kind of restrictions are made and especially, the methods of set theory and analysis are used. Our proof is, then, comparable in this respect to a proof in analytical number theory in which elementary results are likewise won by complicated auxiliary means. Research into foundations [Grundlagenforschung] has to decide on the justification of such a procedure, something that has to be kept separate from sharpness in metamathematics.

We introduce the following inessential modifications to the symbolism of the *Principia*.

1·) We dispense with the ramified type theory and let enter in place of the reducibility axiom the license to substitute in place of function variables arbitrary functions of a suitable type.

2·) We take no signs for variable relations among the basic signs (but only ones for one-place functions), for one can conceive of relations as classes of ordered pairs and ordered pairs again as classes of the second type $\{\ a, b = [(a), (a, b)]$

Analogous holds for relations of a higher type. With this stipulation, one that has in principle no importance anywhere for our proof, type theory assumes the simplest shape.

3·) We replace, something that is very important for what follows, the basic signs of the *Principia* (variables of different types and the logical constants) *in a one-to-one way by natural numbers and the formulas by finite sequences of natural numbers, i.e., functions over finite segments of the number sequence of natural numbers.*

---

[1] [There is a big black bullet drawn right here that points at an addition on the adjacent left page, initially left blank. It is placed here in the way intended, as the next paragraph.]

The formal system laid as a basis presents itself thereafter in the following way:

Basic signs:

0 sign for 0

1 sign for "the successor of"

299L

2 sign for negation

3, 4 square bracket signs (3 = opening 4 = closing bracket)

numbers $\geqq 7$ that are divisible by exactly one prime number are = propositional variables

numbers $\geqq 7$ that are divisible by exactly $k + 2$ different prime numbers are = function variables of type k ($k \geqq 0$)

i.e. variables for a property (class) of kth type

We reserve the numbers **5, 6** to denote the truth values.

We show the concept or through the setting of one next to the other, "all" through the putting ahead of the corresponding variable.

Symbols (= finite sequences of numbers) of the form:

0 10 110 1110 etc are called constant number symbols

and indeed, $\underbrace{11\ldots1}_{k}10$ is called "the symbol for number k"

Improper constant numbers arise when one replaces 0 by a variable of type 0, called variable number symbols.

Let us call an *elementary formula* each propositional variable and further a symbol that consists of a variable of level k followed by a variable of level $k-1$, or of a variable of the first level and a number symbol (constant or variable). (The variables of level $k-1$ and the number symbols are called the arguments of the E 3.)[2]

The operation of "bracketing" of a symbol consists in setting the number 3 in front and the number 4 behind.[3]

---

[2] [E stands possibly for elementary formula, a finite sequence as in point 3 above.]

[3] [A cancelled passage "(obviously understood in the abstract sense)" is followed by a footnote: The expressions often used in the following, "setting in front" "setting next to each other" "bracketing" etc are obviously to be conceived only figuratively.]

The operation of negation **N** of a symbol consists in a bracketing and setting 2 in front.

The operation of disjunction **D** of two symbols consists in a setting next to each other after a preceding bracketing

299R

of each of the two.

The operation of generalization **G** consists in setting a function variable in front, after a preceding bracketing.

What arises from the elementary formulas through arbitrarily repeated application of the operations of negation, disjunction, and generalization we call "formulas," with the restriction that:

1·) **G** can be carried out only on formulas in which the corresponding variable is free (i.e., does not stand between two brackets or in the beginning before a bracket)

2·) **D** can be carried out only on disjoint formulas (i.e., ones in which no bound variable in one is free in the other).[4]

We designate as axioms those formulas, in our sense, that correspond to the following propositions:

1. The last three axioms of Peano.

2. The four propositional axioms of the *Principia* (without the superfluous principle of associativity).

3. The two functional axioms

$$(x)F(x) \rightarrow F(y)$$

$$(x)\ A \vee F(x) \rightarrow A \vee (x)F(x)$$

expressed for all types and all variables.

4. The axiom of choice expressed for all types.

300L

5. The statement of extensionality which says that two functions with

[4] We achieve by the latter stipulation at never designating free and bound variables the same and at never having the scopes of equally designated variables overlap (for we are not dealing with spatial figures but with sequences of natural numbers).

equal ranges [umfangsgleich] are identical.[5]

One has to think of the concept of identity that occurs in the axioms to have been defined in the most comprehensive way possible.

If next formula C is the result of detachment from A and B, then

$B = \mathbf{D}[\mathbf{N}(A), C]$

B is called a result of substitution from A if B arises from A through either:

A propositional variable in A is replaced by a formula distinct from A.

Or:

A function of type k is substituted for a free function variable of type k in A (that stands never in an argument place), in a way that is easily described precisely, i.e., more precisely, a formula F (distinct from A) in which there is singled out a free variable of type $k-1$ as the "empty argument place." (The substitution has to happen so that each elementary component that occurs in x_k is replaced by a formula that arises from F when the empty argument places therein are substituted by the appropriate elementary component.)

300R

The class of provable formulas is the smallest class, closed against the operations of detachment, substitution, generalization, that contains the axioms.

We call a formula without free variables a proposition, one with exactly one variable of type 0 a class sign.

We go now into the exact definition of the concept "true proposition."[6]

We call what are either natural numbers or sets of natural or sets of sets of natural numbers and so on *ad inf* objects and to be precise of respective types $0, 1, 2, \ldots$ *ad inf*. We call finite series of objects (i.e., functions over initial segments of the natural numbers) series of objects [Gegenstandsreihen]. For example, the symbols and formulas defined above are, in particular, series of objects.

[5] Were this proposition not taken among the axioms, it would obviously present in a trivial way an undecidable problem.

[6] The idea of such a definition has been expressed [the word "gleich" cancelled] independently of me by Mr A. Tarski from Warsaw.

We call an elementary formula any series of objects that consists either of just one of the numbers 5, 6 (truth values), or of an object of type k and one of type $k-1$, $k>1$, or of an object of first type and a constant number symbol.

We define the operations of bracketing of negation and disjunction applied to series of objects as above.

We understand by generalization of a series of objects A that which results from A when an object of type k, $k \geqq 1$, is replaced by a variable of type k that does not occur in A, or a constant number symbol is replaced by a variable of type 0, with the expression that arises "bracketed" and an x_k "set ahead."

We call "formula" that which arises from the elementary formulas by the repeated application of the operations **N**, **D**, **G**.

301L

All "propositions" are then, in particular, "formulas."

We call an elementary formula true (**W**) if it consists of only the sign 5, or of two classes of which the second is contained in the first, or of a number symbol and a class of type 1 that contains the corresponding number.[7] We call it false in all other cases. One sees without further ado how one can now define, through recursion relative to the operations **N**, **D**, **G**, the concept of truth for arbitrary formulas and therefore also for propositions.[8]

Now one arrives also quite exactly at proving (through complete induction) that

Each provable proposition is true.

If one substitutes in a formula a for propositional variables truth values

---

[7] [A double arrow indicates that the order of number symbol and class should be reversed.]

[8] [There is an superscript^0 here that should indicate a footnote, then at the bottom of the page a long note with a footnote sign in the form of a thick letter H, possibly drawn over the^0, that continues on the next page. The letter H connects to page 304L that has a very similar text. Gödel's footnote is: One could maintain that the above concept formation "series of objects" is inadmissible because it goes against type theory. Against this it should be remarked that 1. The question is always just of the putting together of finitely many objects. 2. That one can define, quite in accordance with type theory, though not the concept true, still the concept true and at most of level k (where "of level k" means that variables higher than those of type k [don't] occur in the proposition considered), and that this suffices for the proof that follows.]

$[w_1 \dots w_l]$ and for all free function variables objects of the corresponding types [added above: $(x_{t_1} \dots x_{t_k})$], with variables of type 0 not the objects themselves, but the corresponding number symbols, one obtains always a formula that we designate as follows:[9]

$$[a; x_{t_1} \dots x_{t_k} w_1 \dots w_l]$$

If a is especially a class sign and x y natural numbers, then $[a; x]$ denotes a quite determinate proposition.[10]

301R

We call an n-place relation R between objects of the types $t_1 \dots t_n$ *finitely definable* if there is a formula a such that

$$W[a; x_{t_1} \dots x_{t_n}] \sim R(x_{t_1} \dots x_{t_n})$$

One is at once convinced of the following facts:

1·) The relation $x_{n-1} \, \varepsilon \, x_n$ is definable, $f(x_0) \, \varepsilon \, x_1$ as well.

2·) If $R(x_1 \dots x_n)$ and $S(x_1 \dots x_n)$ are definable, then also

a.) $\overline{R(x_1 \dots x_n)}$

b.) $R \vee S$

c.) $(x_1) R(x_1 \dots x_n)$

It follows that all relations that are built up from the ε and the successor relations through the use of $^-, \vee, (\,)$, are finitely definable (these are even precisely the ones that occur within our formal system when one interprets it in terms of content).

Formulas are by our definition objects (namely relations between natural numbers, i.e., classes of type 3 of natural numbers). Therefore many of the concepts defined above that relate to formulas also clearly become finitely definable.

302L

This holds especially for the concepts that concern merely the, so to say, "figurative" properties of formulas.

---

9 [he objects that in $[a; x_{t_1} \dots x_{t_k} w_1 \dots w_l]$]

10 [The y has been heavily crossed out to give $[a; x]$, but the preceding y listed after x left uncancelled, cf. page 302L.]

One gets convinced especially that the following classes (and relations) are finitely definable:

1. provable (*Bew*)

2. class sign

3. a is a symbol for the natural number z

4. $b = [a; x]$ where a denotes a class sign[11]

We order now all symbols through a relation R, after increasing sum of digits in numbers, and with the same sum lexicographically. It leads especially to an arrangement of the expressions for class signs [changed from: relation signs], a counting of the class signs [changed from: relation signs] (that is a one-to-one mapping to natural numbers). We designate this mapping by F so that $F(n)$ is the n-th relation sign, and we note that $y = F(n)$ is a definable relation.

We define now a class K of natural numbers in the following way:

$$K(n) \underset{Def}{=} \overline{Bew[F(n); n]} \qquad 1\cdot)$$

Because the concepts that occur in the definition of K are all definable, then this holds of K as well, i.e., there is a class sign a so that

$$W[a; n] \sim K(n)$$

302R

Because a itself is a class sign, it occurs in the counting F in a determinate position (m), i.e.,

$$a = F(m)$$

We claim now that the proposition $[a; m]$ is undecidable. For

I From $Bew[a; m]$ follows

$$\mathbf{W}[a; m] \text{ so } K(m)$$

Since $\overline{Bew[F(m); m]}$ i.e.,

$$\overline{Bew[a; m]}$$

So we arrive at a contradiction.

---

[11] [Changed from $b = [a; x\,y]$ as on page 301L.]

If we, instead, assume that the negation of $[a; m]$ is provable

II $Bew\,\mathbf{N}[a; m]$ then it follows that

$\mathbf{W}(\mathbf{N}[a; m])$ and from this that

$\overline{\mathbf{W}[a; m]}$ so

$\overline{K}(n)$ i.e., $Bew[F(m); m]$ or

$Bew[a; m]$ and

$\mathbf{W}[a; m]$ that stands in contradiction with $\overline{\mathbf{W}[a; m]}$.

The same contradiction follows already from $\mathbf{W}(\mathbf{N}[a; m])$. We have, then, proved together with the undecidability of $[a; m]$ also $\mathbf{W}[a; m]$, i.e., carried through even a decision on that problem.[12]

303L

One can write the undecidable proposition $[a; n]$, if requested, down *in extenso*. For to obtain a, one needs only to write down the right side of 1 in the symbolism of the *Principia Mathematica*. To determine n, one has just to ascertain that the class sign a in question is in our lexicographical ordering.[13]

Only two properties of the formal system of the *Principia* were essentially used in the above proof, namely

1·) All provable propositions are true.

2·) The concept provable is finitely definable.

The above proof can therefore be applied especially to all systems that arise from the *Principia* through the adjunction of new axioms that fulfil the two conditions 1·), 2·). The finite definability of the class of provable formulas states of course the same as that the class of axioms is finitely definable, and it is naturally sufficient to require this only for the class of the newly added axioms, i.e., we have:

[I] *Each system that arises from the Principia through the adjunction of a class of axioms that is finitely definable and contains no false propositions is not definite*

[12] Even the close connection of our proof to the Richard antinomy stands out. Also the other epistemological antinomies can be reshaped into analogous proofs of the existence of undecidable propositions.

[13] [A seamine-like symbol in the margin indicates a continuation, given on page 305L.]

In particular, all finite classes of axioms are naturally finitely definable, and the same with those infinite ones that arise

303R

through the specification of finitely many axioms for all types. On the whole, all somewhat simple classes of axioms are finitely definable.[14]

As concerns the second of the conditions of the previous theorem, namely that the system must not contain any false (provable) propositions, it cannot be replaced by a requirement of freedom from contradiction, say. For one obtains through the following procedure (V)[15] a definable class of axioms that complements the system $\underline{S}$ into a consistent one that is definite with respect to decision: One adjoins the first undecidable proposition of our lexicographical ordering, then the first proposition undecidable in the extended system, etc, *ad inf*. This system must by the above necessarily contain false propositions.

[Cancellation begins] One can, however, replace the condition that no false propositions are contained by a weaker one that presents a [cancelled: generalization] sharpening of the concept of freedom from contradiction, namely the following:

A. If for a class sign a, $[a;n]$ is provable for each n, then $\mathbf{N}\{Gen_n[a;n]\}$ shall not be provable.

I.e., it shall not be the case that on the one hand, the property a can be shown for each arbitrary number, on the other hand the existence of a number with the property *non a*.

A system that satisfies condition A is called $\aleph_0$-consistent, and we have then the theorem:

II Each system that arises through the adjunction of a definable class of axioms to the system S and that is $\aleph_0$-consistent

---

[14] It follows at once from the above theorem that the class of true propositions is not finitely definable. For otherwise one would obtain, through adjunction [?], a system definite with respect to decision, in contradiction to the above theorem.

[15] [Perhaps for the word Verfahren]

304L

is not definite with respect to decision.

It follows from this, incidentally, at once that there exist systems free from contradiction (even extensions of system S) that, though they are free from contradiction, are not $\aleph_0$-consistent, for the system defined by procedure V offers an example for that. [Cancellation ends. A footnote is indicated, but it remains undecided which. The uncancelled text begins with a big fat H-like symbol used for additions. It connects the paragraph that follows to the very similar footnote on page 301L.]

The concept formation of "series of objects" depends essentially on collecting together *all* the objects that occur in the system S (classes of finite types) into a new domain of individuals, and on arriving at a new hierarchy of types for these, so the theory of types is continued into the transfinite in a sense. To one who would not recognise this concept formation, let it be remarked that, in accordance with the usual theory of types, even if it is not the concept true that lets itself be defined, then instead "true and at most of level n" can be defined for each n (in which a formula of level k means that variables higher than type k don't occur in the formula in question) and that this would suffice for the proof at hand.

304R

[Cancellation begins] The following theorem holds, in addition, for properties definite with respect to decision:

If F is a class sign definite with respect to decision in a system Σ free from contradiction and the proposition $Gen_x F(x)$ [?] undecidable in Σ, then it is true.

For if there exists an x such that $\mathbf{W}(F;x)$, then $(F;x)$ would be provable, and consequently also the corresponding existential proposition (i.e., $\overline{Gen_x F(x)}$).

The property that x is a proof figure for a, that occurs in our proof, is definite with respect to decision, so from the undecidability of the proposition follows: It holds for all x that they are not proof figures for a. [Cancellation ends.]

305L

When one asks by what auxiliary means not contained in the system S this decision was made possible, the answer can be only: through the continuation into the transfinite of the theory of types, used essentially in the extension of the definition of the concept true formula.

One can naturally extend the formalism of the *Principia* so that the transfinite types required for the proof of $[a;m]$ occur in it. Still, one can construct in a quite analogous way a problem unsolvable in the new system (S'), through a lexicographical ordering of class symbols in S'. So one can never come to a system definite with respect to decision along this way, something that hangs essentially together with the fact that there are nondenumerably many types, but only denumerably many basic signs in a closed formal system.

[The seamine-symbol of page 303L is found here.] Consider that the unsolvable problem $[a;m]$ represents, interpreted in terms of content, a proposition of a relatively simple logical structure. For it signifies (compare 1) the inexistence of a proof figure for

305R

a formula $F(n)$ that can be given concretely. The property to be a "proof figure for $F(n)$" is clearly definite with respect to decision. The problem has, then, the character of those like the ones of Goldbach or Fermat: Is there a finite object (number, finite set of numbers, etc) with a property given in advance that is definite with respect to decision?[16]

$$
\begin{aligned}
& fx\,,\ (x)\,,\ ^{-}\,,\ (\imath x) \\
& It(m,n)_{\times}\ \varphi(x)\ = \\
& It(m,0)_{\times}\ \varphi(x)\ = m \\
& It(m,fn)_{\times}\ \varphi(x)\ =\ \varphi[It(m,n)_{\times}\ \varphi(x)] \\
& \quad f[(\imath x)\ \varphi(x)\ \sim \\
& \varphi[(\imath x)\varphi(x)] \vee [(\imath x)\varphi(x) = 0\,\&\,\overline{(y)\varphi(y)}] \\
& \&\,(y)(y < (\imath x)\varphi(x) \rightarrow \overline{\varphi(y)})]
\end{aligned}
$$

[16] [The formulas that follow relate to the developments from page 311R on, perhaps added at some point as the next page 306L begins with the Königsberg lecture, omitted here.]

311R[17]

Z-formula class of numbers

Z-relation *class* *Z-proposition*

W-Z-proposition

[*Z-Rel, x y*] Z-proposition

Bew Z-proposition

Finit. Def

$F(n)$ the n-th class sign

312L [Calculations of binomial coefficients and similar, omitted here.]

312R

$\overline{Bew[F(n),n]} \sim W(K;n)$

$K = F(k)$

$W[K;n] \rightarrow \overline{Bew[F(n),n]}$

(1. Class of *Bew* formulas is *definable*) B

2. $F(n)$ is definable
3. $[a,n]$ is definable
4. class is definable through formula K
5. $\Psi(m,n)$ m is proof for formula $[K;m]$, L
6. L is definite with respect to decision
7. $K = F(k)$
8. W_e of lower level [?] definable for each e W_e
9. E definite with respect to decision and of level n is definable E_e

[17] [The first four lines of this page are the last lines of the Königsberg lecture: I have succeeded, instead, in showing that such a proof of completeness for the extended functional calculus is impossible or in other words, that there are arithmetic problems that cannot be solved by the logical means of the PM even if they can be expressed in this system. These things are, though, still too little worked through to go into more closely here.]

313L

$e > St(K) \qquad E_e = E \qquad W_e = W$

10·) $[E, L]$ is provable

$[W, [L; m, n]] \rightarrow [B, [L; m, n]]$ is provable

11·) $[W, Neg\, E_p(L; p, k)] \rightarrow [K, k]$

$Neg[K, k]$

$Neg[K, k] \rightarrow [B, [F(k), k]]$

$\rightarrow [B, [K, k]]$

$[(B, [K, k])] \rightarrow [W, [K, k]]$

313R

$[K, k] \sim Neg\, E_1[L.\, k]$

$E_1[L.\, k] \rightarrow B\{E_1[L.\, k]\}$

$K = F(k)$

$W[K, n] \sim \overline{Em} Bew[F(n), n]$

$K = Neg\, Ex_1\, L$ —

$Entsch\, L$ —

$W[L; m, n] \rightarrow Bew[L; m, n]$ —

Vor[18] $Bew[K; k]$

$Bew[Neg\, E_1 L; k]$

314L

$[Neg\, E_1 L; k] = Neg[Ex_1 L; k]$

[18] [*Vor* stands for *Voraussetzung* that is best translated as condition here. Gödel's first notes on logic in the notebook *Übungsheft Logik* of 1928–29 contain a system of linear natural deduction in which temporary assumptions are indicated by *Vor*. The next page shows how Gödel's notation is used: The assumption indicated by *Vor* is accompanied by a vertical line drawn until the assumption is closed. In the linear derivation that follows, the assumption leads to a contradiction by which its negation is concluded and the assumption closed at that point. For a detailed explanation, see my "Kurt Gödel's first steps in logic," 2018.]

$$
\begin{array}{l|l}
Vor & W\,[Ex_1 L; k] \\
 & W\,Ex_1[L; .\,k] \\
 & W\,[L; p\,k] \\
 & Bew\,[L; p\,k] \\
 & Bew(Ex_1\,[L; .\,k]) \\
 & Bew[Ex_1 L; k] \\
\end{array}
$$

$$\quad Wid$$

$$\overline{W\,Ex_1[L;k]}$$

$$W(Neg[Ex_1 L; k])$$

$$W[K, k]$$

$$(\overline{Em})Bew\,[F(k), k]$$

$$(\overline{Em})Bew\,[K, k]$$

$$\overline{Bew\,[K, k]}$$

314R

$$\overline{W\,F} \sim W\,NegF$$

$$[NegF; k] = Neg[F, k]$$

$$[Ex_1 L; k] = Ex[L; .\,k]$$

$$Bew[F; n] \to Bew(Ex\,F)$$

$$W(Ex\,F) \sim (Em)W[F; m]$$

Concepts between [?] numbers K, k, L
$[\;,\;]$, W_n, Bew, $F(n)$
Neg Ex_l $Entsch_n$

---

x numerical variable $f(\;)$ successor $\overline{}$ $\vee$ $(\;)$

$x\,\varepsilon(\;)$ the x $[\varphi(1)\,\&\,(n)\varphi(n) \to \varphi(n+1)] \to (n)\varphi(n)$

$xy\,It(\;)mn$ $a + (fb) = f(a+b)$

$(\;)f\,Neg\,\varepsilon\,It\,Id$ $\overline{fn = 1}$ $a + 1 = fa$

$a.b + 1 = ab + a$

315L

1. = one
2. = successor
3., 4. brackets open closed
5. negation
6. *Od*
7. *Id*
8. ε function
9. *Gen*

$FinRe(x)$	$x\,y\,It\,z$	$Var(x)$
$n\,Glv\,x$	$y\,F\,v\,w$	$Zeich(x)$
$x\,Add\,y$	$x\,Frva\,y$	*Bed*
$Re(x)$	$x\,Fre\,y$	
$Nachf(x)$	$x\,Op\,y$	
$Neg(x)$	$x\,Op\,y\,z$	
$x\,Od\,y$	$x\,Op\,u\,v\,w$	
$x\,Gen\,y$	$Form(x)$	
$x\,\varepsilon\,Fy$	$Satz(x)$	
$x\,Id\,y$	$x\,y\,Subst_1\,z$ [19]	

315R

$x\,n\,St\text{-}Eins\,z$

$[z,n]_1 \quad [z,n]_2$

$Gen_1\,y \quad Gen_2 y \;\ldots$

$Satz(x) \sim Form(x)\;\&\;(\overline{Ey})\,y\,Frva\,x$

$Clsz(x) \sim Form(x)\;\&\;(E1y)\,y\,Frva\,x$

---

[19] [Originally $x\,x\,Subst_y, z$. For the index, see page 317L]

$Relz(x) \sim Form(x) \,\&\, (E2y)\, y\, Frva\, x$

$y\, Frva \;\sim Var(y) \,\&\, (En)[n \leqq l(x)) \,\&\, y = n\, Glv\, x]$

$\& \,(\overline{En})[n < l(x)) \,\&\, y = n\, Glv\, x \,\&$

$\& \,((n+1)\, Gl\, x = 9 \vee \ldots = 7 \vee \ldots = 8 \vee (n+2)\, Glx = 8)]$

$1\, Pr\, x = y \varepsilon\, (Prim(y) \,\&\, y/x)$ [*Pr* changed from *Prz*]

$(n+1)\, Pr\, x = y \varepsilon\, (Prim(y) \,\&\, y > n\, Prz\, x \,\&\, y/x)$

$n\, Glv\, x = y \varepsilon\, (x|_y (n\, Prz\, x)^y)$

316L

$Form(x) \sim (Ey)\{x = l(y)\, Glv\, y \,\&$

$(n)(n \leqq l(y) \rightarrow (n\, Glv\, y = 0 \vee Var(n\, Glv\, y) \,\&$

$[(Epqr)\; p, q, r < n \,\&\, (n\, Glv\, y\, Op\, p\, Glv\, y \vee$

$\vee\, n\, Gly\, Op\; pq])$

$l(y) = x\, \varepsilon_x \,\{x\, Gl\, y > 1 \,\&\, (x+1)\, Gl\, y = 1\}$

$x \dotplus y = \varepsilon_z \,\{l(z) = l(x) + l(y) \,\&\, (n)[n \leqq l(x) \rightarrow$

$n\, Gl\, z = n\, Gl\, x \,\&\, l(x) < n \leqq l(z) \rightarrow$

$n\, Gl\, z = \; (n - l(x))\, Gl\, y]\}$

$R(x) = 2^x$

$Einkl(x) = R(3) \dotplus x \dotplus R(4)$

$Neg(x) = R(5) \dotplus Einkl(x)$

$Nachf(x) = R(2) \dotplus Einkl(x)$

$x\, Od\, y = Einkl(x) \dotplus Einkl(y)$

$x\, Gen\, y$

316R

$x\, \varepsilon\, Fy$

$x\, Id\, y$

$xy\, It\, zvw$

$x\, Geb\, y \sim (En)\{n \leqq l(y) \,\&\, n\, Gl\, y = x\}$

$Var(x) \,\&\, \overline{x\, FrVa\, y}$

$x\, Fre\, y \sim (\overline{Ez})\{z\, FrVx \,\&\, z\, Geb\, y \vee$

$z\, FrVy \,\&\, z\, Geb\, x\}$

$Zahl(x) \sim$

$Auss(x) \sim$

$x\, n\, StEins\, y = \varepsilon_z\{(Euv)\, y = u \dotplus R(n\, Gl\, y) \dotplus v$

$\&\, z = u \dotplus x \dotplus v\}$

317L

$x\, y\, Subst_k z$

$x\, y\, Subst_1 z = z$

$x\, y\, Subst_{k+1} z = x\, \varepsilon_n[n\, Gl\, x\, y\, Subst_k z = y]$

$St\, Eins(xy\, Subst_k z)$

$Anz(y, zx) = \varepsilon_k\, [\overline{y\, \varepsilon\, xy\, Subst_k z}]$

$xy\, Subst\, z = xy\, Subst_{Anz(y,z,x)} z$

$[x, y] = y\, FrVa(x)\, Subst\, x$

$FrVa(x) = \varepsilon_y\, \{y\, FrVa\, x\} \,\&\, (En)[n\, Gl\, x = y \,\&\, ((\overline{Em})$

$m < n \,\&\, m\, Gl\, x\, FrVa\, x)]\}$

$Z(1) = 1$

$Z\{n+1\} = Nachf(Z(n))$

$B(x) \sim (n)\{n \leqq l(x) \rightarrow [Ax(n\, Gl\, x) \vee (Epq)$

$p, q < n \,\&\, Folg\{p\, Gl\, x, q\, Gl\, x, n\, Gl\, x\}]$

$\vee\, (Epx)\, p < n \,\&\, x\, FrVa\, p\, Gl\, x \,\&\, n = x\, Gen\, p\, Gl\, x]$

317R

$Folg(xyz) \sim x = (Neg\, y) Od\, z$

$m\, Bew\, n \sim B(m) \,\&\, [l(m)] Gl\, m = n$

$R(1xy) = \varepsilon_n\, \{n\, Gl\, y = x \,\&\, (m)[m > n \rightarrow \overline{m\, Gl\, y = x}]\}$

$R(k+1, xy) = \varepsilon_n \{n\, Gl\, y = x \,\&\, n < R(kxy)$

$\& (m)\{n < m < R(kxy) \rightarrow \overline{m\, Gl\, y = x}\}$

$x\, y\, Subst_1\, z = z$

$x\, y\, Subst_{k+1} z = x, k\, StEins(xy\, Subst_k z)$

$Anz(yz)\, \varepsilon_k[(\overline{En})n < R(kyz)$

$\& n\, Gl\, z = y]$

$Bed\, 1 = 1 \qquad \varphi(0) = 1$

$Bed\, Nachf\, x = f(Bed\, x) \qquad \varphi(n) = 0$

$Bed\, Neg\, x = \varphi\, Bed\, x$

318L

$Bed\, x\, Gen\, y = \varepsilon_z \{(u) Bed[Z(u), x\, Subst\, y] = 1$

$\& z = 1 \vee (u) Bed[\qquad] = 1 \,\&\, z = 2\}$

$Bed\, x\, \varepsilon\, Fy = \varepsilon_z \{(Bed(Z(z), x\, Subst\, y) = 1\}$

$Bed(x\, y\, It\, u\, v\, 1) = Bed\, v$

$Bed(x\, y\, It\, u\, v\, Nachf\, k) = x, Z[Bed(x\, y\, It\, U\, v\, k)]$

$y, Z(k)\, Subst\, u$

$Gr_1(x) \sim (Ey)\{x = Z(y)\}$

$Gr_{n+1}(x) \sim (Ey)\, Op'(xy) \vee (Euv)\, Op'(xuv)$

$\vee (Evwr)\, Op' x(vwr)$

3. The development of mathematics in the direction of greater exactness

318R [The page is very faint at places.]

1. Definiteness with respect to decision

The development of mathematics in the direction of greater exactness has, as is known, led in the end to great parts of mathematics being formalized (as intended, even the whole of mathematics). The presentation to follow treats throughout that kind of "formal deductive systems," the concepts of which we lay down as follows: For a deductive system to be given, the following have to be settled:

1·) What signs (symbols) to count as basic signs?

2·) What combinations of basic signs represent meaningful propositions?

3·) What meaningful propositions to maintain as axioms? If they are finitely many, then to be added through writing them down. If they are infinitely many, then either through giving a [?] property that characterises axioms, or through an *axiom rule*, i.e., a law that assigns to each natural number a meaningful proposition, with the condition that those and only those propositions that are associated to a number are claimed to hold as axioms.

4·) By what rules can one conclude new propositions from the axioms and propositions already proved (rules of inference)?

Delimitation of the concept of a deductive system: The deductive systems to which the following treatise mainly relates are the following:

1·) The system of number theory exactly described on page [323R] (the theory of entire numbers).

2·) The system of *Principia Mathematica* (Here we count among the axioms the axiom of choice

319L [The page is quite weak.]

and the infinity axiom in the sharpened formulation: there exist exactly denumerably many individuals). [added heavily: **Hilb**]

3· The axiom system of von Neumann of set theory (essentially a further development of Zermelo-Fraenkel's)

The three deductive systems introduced stand to each other in an order of one under the other in the sense that all that can be expressed (proved) in a preceding system can also be expressed (proved) in the next one to follow, but not the other way around.

One of the most important questions that can be posed in relation to a formal system is the one about definiteness with respect to decision (completeness in the sense of the Poles), i.e., the question whether each meaningful proposition S from the discipline in question is decidable from the axioms, i.e., either S or *non-S* provable by the rules of inference. In other words, is each problem that is expressible in the formal system under question solvable by the means contained? A very general result is derived in this respect from which it follows that none of the three systems introduced is definite with respect to decision,[1] and that there are in them even undecidable [propositions] of a relatively simple structure, namely, the following theorems hold:

319R

1·) There exist in each of the systems S_1, S_2, S_3 undecidable propositions, and these can also be given, and even infinitely many independent ones in the sense that from no subset of them can the rest be inferred.

2·) There exist properties of natural numbers $F(n)$ definable in S_3 (even more so in each narrower system, e.g., S_2) definite with respect to decision,[2] for which neither $(n)F(n)$ nor $(En)\overline{F(n)}$ is provable.

3·) There exist formulas of the narrower functional calculus for which, in S_2 S_3 (in S_1 no such occur), neither their general validity nor the existence of a counterexample can be proved.[3]

[1] Problem III as posed by Hilbert in his Bologna talk (cf.) will be resolved. Problem IV is already solved in a work of mine [Gödel 1930].

[2] [There is a footnote the end of which is very faint: I.e., they are definable only through recursions and one can therefore give procedures that allow to decide, for each number, whether it has the property F or not.]

[3] [A cancelled paragraph of seven lines with no conclusive sentence structure follows. It mentions an ordinal Ω and ends with: One can complement S_i into a complete system in $\aleph$ different ways.]

It follows from 3 especially that one cannot solve the decision problem of the narrower functional calculus even with the ways of inference of abstract set theory known today.

320L

To be able to express the general result from which the theorems 1–3 follow, let the following be said in advance. The "propositions" "proofs" etc of the deductive disciplines, conceived as concrete figures, are thereby surveyable finite objects (finite series of symbols).[4] One can, however, map finite series of symbols in a unique and isomorphic way on finite series[5] of natural numbers (by replacing the basic signs in a unique way through numbers).

Concepts about series of numbers occur in formalized disciplines themselves (these meant in a contentful way). One can especially define, for example, the concepts "meaningful proposition" and "axiom" of the system S_1 inside this system S_1 (meant in a contentful way) and the analogue holds for systems 2, 3.

The general result from which Theorem 1 follows is the following:

II If a system of deduction satisfies the conditions:[6]

1. The system contains number theory.

2. The class of "axioms" (and the "axiom" rules)[7] are definable within the system (this meant in a contentful way). (This condition will become in particular fulfilled when there are only finitely many axioms at hand).

3. Each "formally provable proposition" is contentfully correct.[8]

---

[4] [A footnote is squeezed at the bottom of the page: When the concepts "proposition" "proof from axioms [?]" "consequence [?]" are to be conceived in a purely formal sense (as properties and relations between series of signs), we put them in quotation marks.]

[5] Natural series in an abstract sense, say, functions over segments of the number series of natural numbers [Belegungen von Abschnitten der natürlichen Zahlenreihe mit natürlichen Zahlen].

[6] [An arrow indicates that the originally third condition has to be placed first. I have adjusted their numbering accordingly, to have a match with references to them on pages 322L and 322R. The cancelled pages that follow contain different formulations of the conditions and the result.]

[7] [Added on the right page: This is especially always the case when the class of axioms is finite.]

[8] We give in what follows an exact definition for this concept, for all cases encountered.

then there are undecidable propositions in it.

320R [cancelled]

321L [cancelled]

321R [cancelled]

322L

Theorems 2 and 3 result through a precise analysis of the proposition that turned out to be undecidable and they hold for all formal systems that satisfy the above conditions 1–3 and have axiom classes (and axiom rules) that are definite with respect to decision (i.e., to be more precise, it must be decidable by number-theoretic means whether a "proposition" is an "axiom" or not, and whether it is the *n*-th axiom).[9]

322R

The above theorem I remains still correct when one replaces condition 2 in it by the sharper one that the axiom rules be definable in the system of number theory (S_1), and condition 3 by the weaker one that each provable proposition of number theory has to be contentfully correct.[10] Even this is of importance, because with this formalization, a contentful interpretation is required only for the system S_1. (For the existence of such for S_2 and S_3 is contested from different directions.)

So as not to let the main idea of the proof of theorem I disappear in the details that follow, it is briefly detailed out here, [?] without the system S satisfying the conditions of theorem I.

If one substitutes for a free numerical variable in a the symbol for the number n, we denote that by $[a; n]$.[11] We think of the class signs from S (of which there are only denumerably many) as lexicographically ordered and call the n-th by $F(n)$, the statement that a is provable in S briefly by $Bew(a)$.

We define now a class K of natural numbers by the stipulation:

[9] This can be expressed also as follows: the axioms must be definable by recursions in system S_1.

[10] Even this condition can be further weakened (cf. p. .)

[11] [This phrase is preceded by a similar inconclusive one that begins with: If in a meaningfully (i.e., by the rules of the grammar) built symbol from S with a free numerical variable]

323L

$$K(n) = \overline{Bew(F(n), n)} \tag{1}$$

It results from conditions 2, 3 that the concepts *Bew*, the ordering relation F, and the substitution relation $[a; b]$ are definable[12] in S, consequently also the class K. We call φ the class sign that corresponds to K. φ occurs in the counting $F(n)$, i.e.,

$$\varphi = F(p) \quad \text{for a determinate } p.$$

It follows from (1) that

$$K(p) \sim \overline{Bew(F(p); p)} \quad \text{i.e.} \tag{2}$$

$$K(p) \sim \overline{Bew(\varphi; p)} \tag{3}$$

The claim is now that the proposition $(\varphi; p)$ that occurs in S is undecidable. Were it provable, then it would be by 3 also contentfully correct,

323R

i.e., $K(p)$ would hold, i.e., because of (3), $\varphi(p)$ would not be provable.

If instead the negation of $[\varphi; p]$ were provable, it would also be correct, so that even $[\varphi; p]$ would be provable, which is impossible. A contradiction follows, then, from both assumptions.

One recognises a close connection of this proof with the antinomy of Richard. Even the other epistemological antinomies can be used for proofs of undecidability.

2

We begin our presentation by describing the formal system of number theory that is fundamental for what is to follow.

[Added between lines: distinction between contentful and formal form]

Basic signs: $1, f, \sim, \vee, =, \Pi, \tau, It$ ( )
$x\ y \ldots z$ *ad inf*

Execution:[13] The mode of usage and translation of these signs is the following:

[12] [This word seems to be lightly cancelled but no alternative written.]

[13] [These are the longhand letters Erl that should stand for Erledigung.]

1 = number 1

$f(x)$ = successor of

$\sim$ = not

$a \vee b$ = a or b

$x = y \quad a = b$ between natural numbers

$x\tau\, F(x)$ the smallest natural number x for which $F(x)$ holds

324L

[Cancelled: The concept for all is expressed through writing ahead the variable in question, for example]

$x\Pi F(x)$ $F(x)$ holds for all x

The symbol *It* replaces the definition by recursion. It is used as in

$x\,y It\{F(xy), m, n\}$

in which $F(xy)$ denotes a two-place [cancelled: function] in the domain of natural numbers the values of which are again natural numbers, m, n instead natural numbers.

$x\,y It\{F(xy), m, n\}$

denotes then the number that the function φ defined by the recursion

$$\left.\begin{aligned} \varphi(1) &= m \\ \varphi(k+1) &= F(k, \varphi(k)) \end{aligned}\right\}$$

assumes for the argument n. There can occur in F, m, n in addition arbitrarily many numerical parameters.

The basic signs $x\,y$... are called "variables." We call a finite, completely arbitrary series of basic signs "series of signs," denoted by a, b, c etc.[14] We define in the domain of series of signs the following operations that give always series of signs from series of signs.

324R

Bracketing of a, i.e., setting the sign (ahead and the sign) after.

[14] Series of signs that consist of the same signs in the same arrangement are to be considered identical.

$O_1(a)$ i.e., setting ahead the sign f, after a foregoing bracketing.

$O_2(a)$ i.e., setting ahead the sign $\sim$, after a foregoing bracketing.

$O_3(a\,b)$ (and $O_4(a\,b)$) setting one after the other of the bracketed a and bracketed b in the serial order a, b, with the writing between of the sign $\vee$ (and =, respectively).

$O^5_\alpha(a)$ (and $O^6_{\alpha\beta}(a)$) ($\alpha\,\beta$ denote two different individual variables).

Bracketing of a and setting before it both of the signs α and Π in the order ($\alpha\Pi$) (and both of the signs $\alpha\,\tau$ in the order α, τ, respectively).

$O^7_{\alpha\beta}(a, b, c)$ ($\alpha\,\beta$ denote some variables).

Bracketing of a, of b, and of c and setting one after the other the bracketed a, b, c (in this order), setting in front the signs α, β, It in the order $\alpha\,\beta\,It$.

325L

We call Z-signs those series of signs for which there are others from which they arise through the application of operations $O^1\,O^4\,O^6$.

We call S-signs those series of signs for which there are others from which they arise through the application of operations O^{2357}.[15]

We define now, for the six operations $O^1...O^6$, corresponding relations between series of signs in the following way:

$\mathfrak{R}_1(a\,b)$ means $a = O^1(b)$ and b is a Z-sign

$\mathfrak{R}_2(a\,b)$ means $a = O^2(b)$ and b is an S-sign

$\mathfrak{R}_3(a\,b\,c)$ means $a = O^3(b\,c)$

$\mathfrak{R}_4(a\,b\,c)$

$\mathfrak{R}_5(a\,b)$ there is a variable such that $\mathfrak{R}_5 = x\Pi b$ b is an S-sign

325R

[Cancelled addition to the previous page]

$\mathfrak{R}_7(a\,b\,c\,d)$

---

[15] This division is justified by the fact that the operations O^1... deliver as results numbers, the latter [?] operations instead propositions.

We define the class of formulas as the smallest class K of series of signs that contains 1 and the variables and is closed with respect to the relations $\mathfrak{R}_1 ... \mathfrak{R}^7$ (in the sense that the first member x of a relation is contained in K if the following ones are).

The concept of a *free* variable is introduced in the usual way.

Formulas without free variables are called normal.

Formulas that are at the same time Z-signs are called number signs, those that are at the same time S-signs are called proposition signs, normal proposition signs briefly meaningful propositions (or even more briefly propositions). We call formulas that arise only through the application of the operation O^1 to 1 or to a variable simple (constant or variable) number signs. Subst!

326L

We go now over to the definition of the class W of contentfully correct propositions.

We declare, to start, through recursion the operation St that associates to each formula a natural number (the level [Stufe] of this formula), in the following way:

$St(1) = 1$

$St(x_1) = 1$ for each variable x_i

$St(O_2 a) = St(a) + 1$

$St(O_3 ab) = St(O_4 ab) = 1 + \max\{St(a), St(b)\}$

$St(O^4 a) = St(O^5 a) = 1 + St(a)$ [16]

$St(O^7 abc) = 1 + \max\{St(a), St(b), St(c)\}$

$St(O^1 a) = St(a)$ when $St(a) = 1$

$St(O^1 a) = 1 + St(a)$ when $St(a) > 1$

We achieve by the last two stipulations that each simple number sign has level 1 (and there are indeed next to them no others of level 1).

Further, the [?] level of a formula does not change when one substitutes for a free variable a constant number sign.

[16] [The first term seems to relate to page 324R before changes were made there.]

326R

We define now a relation *Bed* that associates to each normal formula a natural number (its meaning [Bedeutung]), by recursion according to level:

1. $Bed(1) = 1$

   $Bed\, O^1(a) =$ the number that follows $Bed(a)$

We shall designate by Z the operation, covered by *Bed*, that refers to simple number sign indices,[17] so that hereby *Bed* is defined for every formula of level 1.[18]

2. $Bed(O^2 a) = 1$ when $Bed(a) = 2$ [19]
   $= 2$ when $Bed(a) = 1$

3. $Bed(O^3 ab) = 1$ when $Bed(a) = Bed(b)$
   in contrary case $= 2$

4. $Bed\, O^4(ab) = 1$ when at least one of the two numbers $Bed(a)$ $Bed(b) = 1$
   otherwise $= 2$

327L

$Bed\, O^5_x\, a = 1$ when for each simple constant number sign z
$Bed[Subst(\tbinom{x}{z})a] =$ [20] 1 otherwise $= 2$

$Bed\, O^6_x\, a =$ the smallest number p for which $Bed\{Subst(\tbinom{x}{Z(p)})a\} = 1$ and equal to 1 when no such number p exists[21]

$Bed\, O^7_{\alpha\beta}(a\, b\, c)$ shall [be] the number m that the function $\varphi(x)$ that satisfies the conditions

$\varphi(1) = Bed(b)$

$\varphi(k+1) = Bed\{Subst(\tbinom{\alpha \quad \beta}{Z(k)\; Z(\varphi(k))})a\}$

---

[17] This presupposes that *Bed*( ) is unique for simple number signs, of which it is easy to convince oneself.

[18] [The following continuation is cancelled: The index of the meaning relation for simple constant number signs will be denoted by Z, so $Bed(Za) = a$.]

[19] The numbers 1, 2 represent here the two "truth values," the values true and false.

[20] Observe that $St[Subst(\tbinom{x}{z})a] = St(a)$ or $< St\, O^5_x a$, which gives [?] to each simple number sign z the possibility of recursive definition [?].

[21] [The symbol p is a later addition, with Zp written above some other letter in the substitution.]

assumes for the argument $Bed(c)$ [changed from $Bed(b)$].

One can naturally turn the recursive definitions of St and Bed, through known methods, without further ado into explicit definitions into which we won't go further here.

A proposition S is called contentfully correct when $Bed(S) = 1$.

327R [Two carefully drawn diagonal wavy lines cross this page]

⌈ We introduce the following abbreviations:

$$vu\, It(f(u))(f(x))(z) \underset{Df}{=} x + z \qquad \sim(a)$$

$$(\sim(a)) \vee (b)$$

$$a \rightarrow b$$

$$\sim(x\Pi(\sim(x+z) = (y))) \underset{Df}{=} y > x \qquad \sim((\sim a) \vee (\sim b))$$

We understand by a number-theoretic axiom *each formula* that comes out from one of the following schemes:

$$(\sim(x\Pi(F_x))) \vee F(y)$$

$$F(x\tau(F_x v)) \vee [(x\Pi)(\sim F_x))\,\&\,(x\tau(F_x)) = e]$$

$$y < (x\tau(F_x)) \rightarrow \sim F_y$$

$$uv\, It\{F_{uv}xe\} = x$$

$$uv\, It\{F_{uv}x\, f(y)\} = F(y, uv\, It\{F_{uv}xy\})$$

$$(F(e)\,\&\, x\Pi(F(x) \rightarrow F(f(x)))) \rightarrow x\Pi F(x)$$ ⌋

328L [A page of seemingly unrelated computations, written upside down.]

328R

a is called an immediate consequence of b when a and b are proposition signs and there is a variable v so that $a = O^4_v b$.

Further, a is called an immediate consequence of b and c when all three are proposition signs and when

$$b = O(Oc, a)$$

A proof is a finite series of formulas of which each is either an axiom or an immediate consequence of a preceding one (or of two). A formula is called provable when it is the endformula of a proof. It is quite easy to show exactly that each provable formula has the property by which the substitution of simple number signs for the free variables that occur in it results in a contentfully correct formula (then especially, every provable proposition is contentfully correct). The proof proceeds by complete induction in which one shows that each axiom is contentfully correct and that this property is carried over to immediate consequences.

We map now the series of signs to finite sequences of natural numbers through a unique relation R, by letting natural numbers correspond to the basic signs in the following way:[22]

329L

[Lightly cancelled: We map the finite sequences of natural numbers further into the natural numbers, through a relation S that is defined as follows:

The relation S holds between the sequence $x_1 \ldots x_n$ and the number y if and only if

$$y = p_1^{x_1} p_2^{x_2} \ldots p_n^{x_n}$$

Here the p_i are the prime numbers contained in y and $p_1 < p_{i+1}$.]

[Clearly cancelled: For one can map finite sequences of natural numbers on natural numbers themselves, by having each natural number of the sequence as an exponent in the representation as a product of powers of prime numbers.]

In this way, to each natural number is assigned one and only one sequence of natural numbers, and one obtains in this way each finite sequence of natural numbers.

[Lightly cancelled: By this mapping of formulas to natural numbers, all metamathematical concepts defined so far that concern the system S, go

---

22 [Omitted but with some space left to fill in, likely as on pp. 298R and 299L:

0 for 0

1 for successor

2 for negation

3, 4 for (and)

5, 6 the truth values.

numbers $\geqq 7$ divisible by exactly one prime number = propositional variables

numbers $\geqq 7$ divisible by $k+2$ different prime numbers = function variables of type k]

over into properties and relations between natural numbers.]

329R

One can now place on the side of each of the metamathematical concepts defined above a corresponding concept about natural numbers, by the stipulation that this concept shall hold for numbers if and only if the corresponding metamathematical concept applies to the (uniquely) corresponding series of signs.[23] The association between numbers and series of signs is unique only in one direction and therefore one has still to adjoin in the case of operations (for example): one has to take always the smallest number that is associated to the corresponding series of signs.

When in what follows expressions so far defined are used, the corresponding concepts about natural numbers are meant all along. The contrary case is indicated by setting a *B* ahead, for example (*B* formula).

We define now: a class of natural numbers is called definable if there is a sentence-formula[24] a in which at most the variable α is free, such that

$$(x)W\{Subst(\tbinom{\alpha}{Zx})a\} \sim K(x)$$

330L

Analogous definitions hold for relations (with arbitrarily many places) between natural numbers. We put up analogous definitions for operations within the domain of natural numbers.

$\varphi(x)$ is finitely definable when there exists a number formula a with the variable v

$$(x)\{Bed\, Subst(\tbinom{v}{x})a = \varphi(x)\}$$

It is easy to prove exactly that [cancellation to page 330R begins] if the relations $R(x_1 \ldots x_n)$, $S(x_1 \ldots x_n)$ are finitely definable, then also

$$R \vee S \quad R \,\&\, S \quad \overline{R} \quad (x_1)R(x_1 \ldots x_n) \quad (Ex_1)R(x_1 \ldots x_n)$$

Further also the $n-1$ place operation:

The smallest x_1 for which $R(x_1 \ldots x_n)$ or 1 if there is no such x_1.

[23] The concepts *Bed*, *St*, *Z*, to begin with, already relate to natural numbers.

[24] One can give *Z* definable classes, for example, the class of true propositions is one, as will turn out later.

It is equally easy to see that if the operations $\varphi(x_1 \ldots x_n)$, $\psi(x_1 \ldots x_m)$ are definable, then also

$$\varphi(\psi(z_1 \ldots z_m) x_2 \ldots x_n)$$

Further, the relation

330R

$$\varphi(x_1 \ldots x_n) = \psi(y_1 \ldots y_m)$$

and finally those functions that satisfy the conditions:

$$f(1\, x_1 \ldots x_n) = \varphi(x_1 \ldots x_n)$$
$$f(k+1\, x_1 \ldots x_n) = g\{x_1 \ldots x_n\, k\, f(k\, x_1 \ldots x_n)\}$$

[cancellation from page 330L ends]

It is easy to prove exactly on the basis of the foregoing definitions:

S[25]

When the relations $R($ $\quad$ S

and the operations $\quad$ are Z definable, then

also the relations

and the operations

331L

and the operations $\quad$ that satisfy the conditions

One can now show, on the basis of this theorem as well as the fact that the operation of successor (through B formula $f(x)$) and the relation of identity (B formula $(x) = (y)$) are definable, that all of the metamathematical concepts introduced earlier (as listed above), (with the exception of *Bed* and

[25] [Most likely by **S** is meant a theorem (Satz). The result is just indicated and space left here and there so that it can be filled in later in the way suggested by the preceding long cancelled passage that begins identically.]

W), are Z definable. The proof procedure is in short: To show that a relation (class, operation) is Z definable, one gives a series of relations (classes, operations) $\varphi_1\ \varphi_2 \dots \varphi_n$ that begins with *Iden* and the successor operation and ends with R, each φ_n of which arises from the earlier ones through application of the logical operations mentioned in *theorem* .

331R

We assume for the following that the operations of addition $a + b$, multiplication $a \cdot b$, and exponentiation a^b are finitely definable. We step over the proof of this.

In the following, a series of number classes (relations, operations)

$F_1\, F_2 \dots G_1 \dots$

is defined, and indeed defined in such a way each from the previous or the operations ($a + b\ a \cdot b\ a^b$) that theorem VI is each time applicable (several steps of theorem 6 are almost always put together into one here). By theorem VI, all classes (all operations) defined in what follows are Z definable, and among them occur in 6 all of the above listed.

332L [blank]

332R [The page begins halfway down]

We are now in the position to prove the following theorem:

The formal system T [cancelled: that contains Z] satisfies the following conditions:

1. Each meaningful proposition in Z is also a meaningful proposition in T.

2. The axiom class of T (i.e., more precisely that class of numbers that arises from T after the replacement of the basic signs by numbers) is Z definable.

3. Each proposition of Z provable in T is contentfully correct.

Then there exists a proposition A from Z for which neither A nor $\overline{A}$ is provable in T.

333L

The system Z itself, especially, satisfies the above three conditions.

333R [blank]

334L

In the formalistic conception, one adds to the (meaningful) propositions [Sätze] of mathematics transfinite statements [Aussagen] that have no meaning in themselves, but serve only to make the system into a rounded one, just as in geometry, one arrives at a rounded system through the introduction of points at infinity. This conception presupposes that when one adds to the system S of meaningful propositions the system A of transfinite propositions and then proves a proposition from S through a detour over propositions from T [changed from A], this proposition is also correct, so that no false (meaningful) propositions become provable through the addition of transfinite axioms. Hilbert has replaced this requirement by the requirement of freedom from contradiction. One has so far paid little attention to the fact that these two formulations are in no way equivalent.

334R

For if one has proved a meaningful proposition F in a formal system A with the help of transfinite axioms, it follows just that non-F is not provable *within the system A*. Nevertheless, there could be a proof for non-F that is contentfully correct [added above: satisfies all intuitionistic requirements] but is *not* representable in system A. It would be, for example, thinkable that one proves a proposition of the form $(En)F(n)$ (where F is a finite property of natural numbers) by the transfinite means of classical mathematics, and could on the other hand have by contentful finite consideration the insight that all numbers have the property non-F, and such could be even thinkable had one proved the freedom from contradiction of the contemporary system of classical mathematics. For it is altogether not sure whether each contentful consideration can be represented in this system (say ramified M [?])[26]

One can even give examples of propositions (in fact, of the kind of Fer-

[26] [The parenthetical addition with "ramifiziert" written in ink could stand for a system of ramified set theory.]

mat) that one can recognise as correct through contentful consideration but that are undecidable in the system of classical mathematics. Therefore, if one adds the negation of such a proposition to the axioms of classical mathematics (like *PM*), one obtains a system free from contradiction in which a proposition is provable that one can recognise as false through unobjectionable contentful considerations.

335L [This page contains seemingly unrelated computational formulas.]

335R [This page contains unsystematic recursion equations and attempts at explicit definitions of functions, perhaps in relation to the remark on page 326L about turning recursive definitions into explicit ones.]

$\Phi(0,x) = \Psi(x)$

$\Phi(n+1,x) = M(n,\varphi(n,x)\,x)$

$\varphi(y,0,x) \sim \psi(y,x)$

$\varphi(y,n+1,x) \sim$

$\quad (Ez)\mu(y,n,z,x) \,\&\, \varphi(z,n,x)$

$(E\varphi)\{[\varphi(y,0,x) \sim \psi(y,x) \,\&$

$\quad (y)[y \leq n \rightarrow \varphi(y,n+1,x) \sim$

$\qquad (Ez)\mu \quad \& \quad \varphi \quad]\}$

$\varphi(y,k,x) \sim (Ef)\{Eind\, f \,\&$

$[0fy \rightarrow \psi(y,x)] \,\&\, (y,p,n)[n < k \rightarrow nfy \,\&\, (n+1)fp$

$\rightarrow \mu(p,n,y,x)] \,\&\, kfy\}$

$(x)x_2(x) \sim f(x) \rightarrow (x)x_2(x) \qquad (x)x_2 \sim f \rightarrow x_2 \sim f$

$(Ex)x_2(x) \sim f(x) \qquad \rightarrow x_2(x) \rightarrow f(x) \,\&\, f(x+1) \rightarrow x_2(x+1)$

$\quad (x)[f(x) \rightarrow f(x+1)]$

4. The question whether each mathematical problem is solvable

339R[1]

The question whether each mathematical problem is solvable, i.e., whether for each mathematical proposition *A* either *A* or non *A* is provable, lacked so far an exact sense, as the words "mathematical proposition" and "mathematically provable" had not been made precise. The divergence of opinion of various mathematicians on this point is proved sufficiently by the discussions on the axiom of choice and the law of excluded middle. The way to make precise the concepts of "mathematical proposition" "mathematical proof" that lies at the base of the following investigation is essentially the one given in *Principia Mathematica*. More precisely: We take the Peano axioms with the logic of the *Principia* as a superstructure. There are in the *Principia* itself natural[ly?] unsolvable problems, specifically the question of the number of individuals. As is known, all so far known theorems of all mathematical disciplines can be proved in this system, with the exception of certain theorems of the abstract theory of sets that deal with "aleph" cardinalities. So, the appearance is as if really all thinkable mathematical proofs, at least the disciplines that don't treat great cardinalities (number theory, algebra, function theory), were contained in it, and nevertheless, as will be shown, undecidable questions [cancelled: problems] can be given in this system,[2] and even problems of a relatively simple kind, namely questions about the existence and nonexistence of finite sequences of natural numbers with properties given in advance (even definite with respect to decision), and these problems are not decidable even with the help of the axiom of choice.

340L

The method that leads to this result is most suited to generalization. It leads especially to a situation in which one cannot make the *Principia* into a system definite with respect to decision through the adjunction of finitely many new axioms (and, say, such infinite ones that arise from it through type elevation), and the same holds also for all somewhat "simple" infinite ex-

[1] [Three additions are indicated at the facing left inside cover page 338, placed as intended in the following. These two pages have been exposed twice and appear therefore on frames 338 and 339.]

[2] I.e., problems that can be formalized in the *Principia Mathematica* but not solved.

tensions of the *Principia*. Here one requires always of an extension that no false propositions about natural numbers are provable in it, or that they are free from contradiction in a sense to be given below. Proof methods [line ends]

340R

We give next a precise metamathematical description of the formal system for which we want to prove that there exist undecidable problems in it, a system that coincides up to inessential points with the one of the *Principia Mathematica,* as such a description is not to be found in the *Principia Mathematica* itself. We make now, in advance, the following remarks about it.

1. We dispense with the ramified theory of types, therefore allow that each function of a suitable type can be substituted for a function variable, irrespective of whether there appear in it bound variables of higher types or not. It is easy to convince oneself that this stipulation is equivalent with keeping the ramified type theory and the reducibility axiom (that is the standpoint assumed in the *Principia*).[3]

341L [cancelled]

341R

2. We dispense with the introduction of relational variables among the basic signs: For one can conceive of each relation as a class of ordered pairs and each ordered pair as a class of second type, e.g., $a, b = [(a), (a, b)]$, so each relation between individuals as a class of third type. It is easy to ascertain that the analogue holds also for the relations of higher types and inhomogeneous relations. This arrangement built into the *Principia Mathematica* is, incidentally, not at all essential for the proof that follows, but fairly said serves to simplify it.

3. We take among the basic signs a relation f between individuals for which we postulate the Peano axioms (f as the successor relation). So, our system is strictly speaking the axiom system of Peano with the logic of the *Principia Mathematica* as a superstructure. Even this measure serves, fairly said, the simplification of the proof.

---

3 [The end of this page and the next contain the cancelled item 2, rewritten on page 341R.]

4. We add as an axiom the proposition that all functions are extensional (exact formula see page), because otherwise the question whether all functions are extensional would obviously be in a trivial way an unsolvable problem.

342L

The formal system laid as a foundation presents itself thereafter as follows:

1. Basic signs

$\vee$ or[4] $\sim$ negation	
0 = zero f = successor	used in the combination fx, the number that follows x
$p_1\, p_2 \ldots p_k \ldots$ *ad inf*	propositional variables
$x_1^0\, x_2^0 \ldots x_k^0 \ldots$ *ad inf*	individual variables
$x_1^1 \ldots \ldots x_k^1 \ldots$	variables for classes of the first level, simple execution
( )[5]	brackets

Symbols for or and generalization are superfluous, because they are defined from these in the known way through abbreviation;

2.

We understand by a *"series of signs"* each finite series that consists of the above basic signs.

We understand by a *"sum of two series of signs" a b* the series of signs that arises

342R

when first a and then b immediately following is written down.[6]

We understand by "an f series" a series of signs that consists of only signs f.

A *"constant number symbol"* is a sum of an f series and of 0.

[4] [This connective has been cancelled.]

[5] [These have been cancelled.]

[6] [The next three lines are cancelled and the rest as well as pages 343L and 343R crossed over by a diagonal line, but the latter are included here for a better understanding of what follows.]

A *"variable number symbol"* is a sum of an f series and a variable of type 0.

An *"elementary formula"* is the sum of a variable of level $k+1$ and a variable of level k or of a variable of level 1 and a number symbol {so we express the ε relation by the simple writing of one next to the other}, or a propositional variable.

We understand by *"negation of the series of signs a"* that which arises from a when one brackets it (i.e., sets a sign (ahead and a sign) after) and then the sign $\sim$ in front. $Neg\, a = \sim(a)$

343L

We understand by a *"disjunction of the series of signs a and b"* the series of signs that arises when one brackets a and b, then adds them up, $Disj(ab) = (a)(b)$

We understand by *"the generalization of the series of signs a by means of the variable z"* the sum of z and the bracketed series of signs a.

$$Gen(za) = z(a)$$

A variable *"x is a bound variable of the series of signs a"* when it is at some place in a between two bracket signs, or at the beginning and before a bracket.

"x is free variable in a" means that x occurs in a but is not a bound variable in a.

We say of a generalization *"a is an allowed generalization of b,"* the meaning being that the generalization variable occurs in a as a free variable.

We call c an allowed disjunction of a and b when c is a disjunction of a and b and when none of the bound variables from a occur in b and none of the bound variables from b in a (*disjoint formulas!*).

A formula. We can now define the concept "formula." The class of formulas is the smallest set of series of signs that contains all the elementary formulas and is closed with respect to negation, allowed disjunction, and allowed generalization.

343R

We have achieved by our stipulation on allowed disjunction especially that different bound variables, i.e., ones the scopes of which are different, are designated differently in each formula, and that no free variable is equal to a bound one.

We say of a formula a that it *arises through k-fold type elevation* from b if it arises so that one replaces in b each variable of type m by one of type $m+k$, and to be precise different by different, and propositional variables again by propositional variables (for type elevation to be possible, the sign f must not occur in b). We say in place of 0-fold type elevation also congruence.

To be able to write down the axioms more easily, we introduce the following abbreviations:

344L

$$(\sim(a))(b) \qquad a \rightarrow b$$

in place of[7] $x^1\ x^1 x^{0^\bullet}$

$$\sim((\sim(a))(\sim(b)))^\bullet \qquad a \,\&\, b$$

$$a \equiv b$$

$$x^1(x^1 x_1^0 \equiv x^1 x_2^0) \qquad x_1^0 = x_2^0$$

1. Propositional axioms 2. Functional axioms

3. Peano axioms 4. Extensionality axiom

5. Axiom of choice

$$\sim(fx = 0)$$
$$(fx = fy) \rightarrow (x = y)$$
$$((x^1 0) \,\&\, (x((x^1 x) \rightarrow (x^1 fx)))) \rightarrow (x^1 y)$$
$$(x((x^1 x) \equiv (y^1 x))) \rightarrow (x^1 = y^1)$$

344R

Symbols of the form

[7] [This seems incomplete.]

$0\,,\ f0\,,\ ff0\,,\ fff0$ etc

and those that emerge from these when one replaces 0 by a variable of type 0 are called *number* symbols and counted together with the variables of type 0 as *objects of type* 0.

We call an *elementary formula* a combination of signs that arises through the replacement of a place holder in a variable of type $k+1$ by an object of type k. That which appears through a finite repetition of the operations $(\ )\ \vee\ ^{-}$ on elementary formulas is called a *formula*. A formula without propositional and free variables is called a proposition.

We designate as *axioms* the following formulas:

I Propositional axioms

Functional axioms

II 1. $(x)x_1(x) \rightarrow x_1(y)$ and all those that arise from type elevation and renaming of variables

2. $(x)\ A \vee x_1(x) \rightarrow A \vee (x)x_1(x)$

Extensionality axiom

III $(x)[x_1(x) \sim y_1(x)] \rightarrow x_1 = y_1$ and type elevation

Choice

IV $\{(x_1)[x_2(x_1) \rightarrow (Ex)x_1(x)]\ \&$

$(x_1 y_1)[x_2(x_1)\ \&\ x_2(y_1) \rightarrow \overline{(Ex)x_1(x)\ \&\ y_1(x)}]\}$

$\rightarrow (Ex_1)[(y_1)(x_2(y_1) \rightarrow (Ex)(y)\{y_1(x)\ \&\ x_1(x) \sim x = y\})]$

345L

Peano

V. 1. $\overline{fx = 0}$

2. $fx = fy \rightarrow x = y$

3. $x_1(0)\ \&\ (x)[x_1(x) \rightarrow x_1 fx] \rightarrow (x)x_1(x)$

Rules of inference

1. Rule of detachment: from $A \rightarrow B$ and A, B can be inferred.

2. In whatever expression A, any expression distinct from A can be substituted for a propositional variable.

3. If A is an expression and if the free variable x_k of level k occurs nowhere in A as an argument, i.e., x_k must stand in the first place in all elementary components in which it occurs, and if B is any expression distinct from A that contains the free variable y_{k-1} of level $k-1$, then it is allowed to put B in the place of all the elementary components that contain x_k and occur in A, wherein the variable y_{k-1} must be replaced each time by the respective argument of x_k.

4. Arbitrary number symbols can be substituted for variables of type 0.

5. If A contains the free variable x (of an arbitrary type), then $(x)A(x)$ is allowed[8]

6. $\imath$ denotes the variables.

345R

to be inferred from it.

A formula is called provable if it can be obtained from the axioms by finitely many applications of the rules of inference, and the theorem to be justified can itself be expressed in the following way: There are propositions for which neither A nor $\overline{A}$ is provable.

In 3, we have to assume about x_k that it does not occur in any argument place, because no formula in our sense would result through the substitution of an expression in an argument place. This, however, means altogether no restriction in coverage for axiom 3, for it is easy to convince oneself that all propositions about classes of the *Principia* that are provable with the help of the reducibility axiom, follow from our axioms I–IV. The same holds for the propositions about relations, as soon as these are introduced in way given above, page .

[8] [Point 6 has been squeezed at the bottom of the page after the last words of point 5 had already been continued on the next page. The first sign looks just like the description operator $\imath$.]

346L[9] [*Anzeiger* manuscript, first page]

If one builds on top of the Peano axioms the logic of the *Principia Mathematica* (numbers as individuals), with the axiom of choice for all types, a formal system S arises for which the following theorems hold:

I·) The system S is not definite with respect to decision i.e., there exist within it propositions (such can even be given) for which neither A nor $\overline{A}$ is provable, namely, there exist some undecidable problems of the simple structure $(Ex)F(x)$, in which x runs over the natural numbers and F is a property of natural numbers definite with respect to decision. Problems of this kind [?] their unsolvability at the same resolved (in a metamathematical way).

II·) Even if one allows all the logical means of the P.M. (especially the extended functional calculus and the axiom of choice) in metamathematics, there is no

346R [*Anzeiger* manuscript, second page]

proof of freedom from contradiction for system S (even less so if the methods of proof are somehow restricted). So, a proof of freedom from contradiction for the system S can be carried through only by means that lie beyond the system S, and the case is analogous also for other formal systems, say Zermelo Fraenkel's axiom system of set theory.

3·)[10] The Roman I can be sharpened in the sense that the addition of finitely many axioms to the system S (or infinitely many ones that come out from them through "type elevation") does not lead to a system definite with respect to decision, as soon as the extended system is $\aleph_0$-consistent.

Here a system is called $\aleph_0$-consistent, when for no property of natural numbers F, $F(1)$ $F(2)$ $---$ $F(n)$ *ad inf.* are at the same time provable, and instead $\overline{(Ex)F(x)}$ { there are extensions of the system that are free from contradiction but not $\aleph_0$-consistent}.

4·) I continues to hold also for all extensions of the system S through infi-

[9][The writing on this page is weak, as if some lines had been erased. The typewritten version follows closely these two handwritten pages and provided decisive help in their reading.]

[10] [Results I and II on the previous page had first been 1 and 2, then changed, but not 3 and 4 on this page.]

nitely many axioms as soon as the class of axioms K added is definite with respect to decision. It is metamathematically decidable for each formula if it is an axiom or not (here again, the logical means of the *Principia* are assumed in metamathematics).

347L [A blank page.]

347R

We have so far defined a series of concepts that relate to series of signs, without encountering the question of what these signs (and series of signs) actually are (whether formal or physical objects etc). [Cancelled: If an answer had to be given to that, it could very well not be other than that the signs are physical objects. Hereby metamathematics would become a part of physics and even that is acceptable. The solution to this difficulty lies at hand:] It is obviously quite indifferent for metamathematical investigations what one lays down as signs to serve as a basis. The question is only about their equality and difference and their arrangement in the formulas, but even this arrangement need not be in any way anything spatial. We can take, for example, without further ado as signs natural numbers and as series of signs finite sequences of natural numbers, i.e., number-theoretic functions defined on segments of the number series. The above definitions are, then, to be completed with something like the following ones:

basic sign = natural number

propositional variable = number that is divisible by exactly one prime number

variable of level k number that is divisible by exactly $k+2$ prime numbers

formula = finite sequence of natural numbers (that satisfies certain conditions)

When one imagines further the sign 0 to be replaced through the number 0, f through 1, the brackets through 4 5, [cancelled but intermittently underlined for emphasis: one sees easily how all of the concepts defined above

348L

go over into properties and relations of natural numbers.]

Formulas are, then, from this point of view finite sequences of numbers with certain properties and metamathematics a science that treats certain properties of finite sequences of numbers. We gain two possibilities through this conception.

1. It clearly follows that metamathematics holds in no way any exceptional position against other mathematical disciplines, for it treats as said of finite sequences of numbers and these appear in the most different parts of mathematics. It follows especially that no restrictions whatsoever are required in the proof methods for the proofs of metamathematical theorems. I.e., one can apply all theorems and methods of analysis and set theory etc in metamathematical proofs. A proof of a metamathematical theorem thus conducted is comparable to proofs in analytical number theory. Here as well as there, simple combinatorial results (theorems about finite numbers or

348R

finite sequences of numbers) are won by higher auxiliary means. (It is another question to what extent such a procedure is justified on the whole (in analytical number theory as well as metamathematics), one that hangs together with the proof of freedom from contradiction and belongs to foundational research that is to be sharply separated from metamathematics).

2. From our conception results the strange state of affairs that the metamathematical theorems (or at least a part of them), because they are theorems about sequences of numbers, can be expressed in the system itself, the metamathematics of which one exercises (in the case at hand, *Principia Mathematica*). It is to be expected from the outset that such expressibility must yield interesting results, and in fact, the whole of the proof to follow depends on it.

349L

We begin now the systematic presentation of the proof and take here the point of view by which combinations of signs are finite sequences of natural numbers. Hereby all metamathematical concepts explained in the following receive a completely different sense from the one one is otherwise used to have with them. The reader is begged to forget this usual sense as far as possible and instead to understand by these concepts solely that which is expressed in the definitions that follow.

Definition 1.* x_3 is a series of signs $\underset{Df}{=}$

x_3 is a function over a segment of the series of natural numbers to natural numbers $\geqq 0$.[11]

We designate by $x_3(n)$ the number that stands in the n-th place. We designate a number series of length 1 for which $x_3(1) = a$ with $\bar{a}$. We designate series of numbers by x_3, y_3, numbers by x, y.

Definition 2.* x is the length of the series of signs $x_3 =$

x is the greatest natural number for which there is still assigned a number in x_3 $l(x_3)$.

Definition 3.$^\times$ x_3 is the sum of series of signs y_3 and $z_3 =$ $x_3 = y_3 + z_3$

$x_3(k) = y_3(k)$ for $k \leqq l(y_3)$

$x_3(k) = z_3(k)$ for $k > l(y_3),\ k \leqq l(y_3) + l(z_3)$[12]

Definition 4. x_3 is a propositional variable =

349R

[cancelled: x_3 has length 1 and $x_3(1)$ is a number] > [positive] that is divisible by exactly one prime number.

Definition 5. x_3 is a variable of level k = $(k \geqq 0)$

is > [positive] and divisible by exactly $k+2$ prime numbers.

Definition 5a. x_3 is a variable $x_3 >$ and by at least[13]

Definition 6. "x_3 is a constant number symbol" =

let n be the length of x_3 then the following must hold:

$x_3(k) = 1$ for $k < n$ $x_3(n) = 0$[14]

Especially the series that consists of just 0.

Definition 7. "x_3 designates the number x" is a symbol for number x

---

[11] [The German is: x_3 ist eine Belegung eines Abschnittes der Reihe der natürlichen Zahlen mit Zahlen $\geqq 0$.]

[12] [The original has $x_3(k) = z_3(k-m)$, from an unfinished cancellation with m the length of y_3 and $x_3(k) = z_3(k-m)$ for $k > m$.]

[13] [This definition has been squeezed between the lines and the last words "2 prime numbers" cancelled.]

[14] [Original has as a leftover from a second cancelled alternative: either = 0.]

x_3 is a constant number symbol $\quad l(x_3) = x + 1$

Definition 8. "x_3 is a variable number symbol"

$x_3(k) = 1$ for $k < l(x_3)$

$x_3(k) =$ divisible by exactly two prime numbers for $k = l(x_3)$

Every variable of type 0 is in particular a variable number symbol.

"x_3 is a number symbol" = x_3 is a constant or variable number symbol.

350L

Definition 9. We designate the sequence of signs $\overline{2}$, $\overline{3}$, and $\overline{4}$ as, respectively, negation symbol, open bracket sign, closed bracket sign.

Definition 10. x_3 is a negation of $y_3 \quad x_3 = N(y_3)$

$x_3 = \overline{2} + \overline{3} + y_3 + \overline{4}$

[Cancelled: Definition 11. x_3 is a disjunction of y_3 and $z_3 \quad x_3 = y_3 O z_3$

$x_3 = \overline{3} + y_3 + \overline{4} + \overline{3} + z_3 + \overline{4}$]

Definition 10.

"x_3 is an elementary formula"

x_3 is either a sum of a variable of level $k+1$ and level k or of a variable of level k and a number symbol.

We designate in what follows all by simply setting the corresponding variable before the bracketed formula, the scope of which it is. Therefore we can define bound variables simply by their standing either at the head of a formula or between two brackets.

Definition 11. "x is a bound variable in x_3"

there exists n so that $x = x_3(n)$ and $x_3(n-1)$ as well as $x_3(n+1)$ is a bracket sign

or $x = x_3(1)$ and $x_3(2)$ is a bracket sign.

350R

Definition 12. "x is a free variable of x_3"

x occurs in x_3 but is not a bound variable of x_3.

Definition 13. "x_3 is distinct from y_3"

no bound variable of y_3 is free in x_3 and no bound variable from x_3 is free in y_3.

Definition 14. "x_3 is is the disjunction of y_3 and z_3"

$x_3 = \overline{3} + y_3 + \overline{4} + \overline{3} + z_3 + \overline{4}$

and y_3 is disjoint from z_3.

Definition 15. y_3 is the generalization of x_3 with respect to the variable x

x a free variable in x_3 (not a propositional variable) and

$y_3 = \overline{x} + \overline{3} + x_3 + \overline{4}$

~~Definition 16~~. y_3 is generalization of x_3

there exists a variable such that y_3 is a generalization of x_3 in relation to this variable.

Definition 16. "y_3 is a formula (or expression)"

x_3 belongs to the smallest set of series of signs that contains the elementary formulas and is closed

351L

against negation, generalization, and disjunction.

We have achieved through our arrangement about the disjointness of formulas, the disjunction of which is built, that the scopes of bound variables identically denoted never overlap, and that no bound variable is designated by a free one equal to it.

Definition 17. x_3 is a proposition x_3 is a formula without free variables.

Definition 18. x_3 is a two-place relation sign

it contains exactly two different free variables that are both of type 0.

We go now over to the exact definition of the class of formulas one usually describes with the words "true proposition," by which one wants to express that the thought expressed through it is true. It is one of the cardinal points of our proof that this concept is amenable to an exact mathematical formulation in which one doesn't have to invoke in any way a mysterious "meaning relation," but uses only concepts otherwise usual in mathematics. We introduce to this purpose the concept:

Definition 19. "Series of sets." We understand by such a function over a segment of the series of natural numbers to natural numbers, or of sets of numbers, or of sets of sets of numbers and so on, whereby sets of arbitrarily high types are allowed to appear, and in the series of sets also

351R

sets of different types can occur. The series of signs defined above are contained therein as special cases. This concept formation falls beyond the Russellian type theory by the consideration that there occur only finitely many sets in each such series, and therefore also a finite highest type must occur, whereas one could hardly object anything to that.

One can explain just as above the sum of two series of sets.

Definition 20. We shall call a series of sets f of length 1 for which $f(1)$ represents a class of type k an "element of type k."

Definition 21. We understand by an "elementary formula" a series of sets that is either the sum of an element of type $k+1$ and one of k or the sum of an element of type 1 and a constant number symbol or finally identical with $\overline{5}$ or $\overline{6}$ (the numerical values 5 and 6 represent for us the truth values true and false).

Definition 22. We understand by the negation of formula f the formula

$$y = \overline{2} + \overline{3} + f + \overline{4}$$

Definition 23. We understand by the disjunction of the formulas f and g the formula

$$h = \overline{3} + f + \overline{4} + \overline{3} + g + \overline{4}$$

352L

Definition 24. We understand by the generalization of the series of sets f the following:

One replaces a set of type k that occurs in f by a variable of type k that does not occur in it (= number that is divisible by $k+2$ prime numbers). One builds from the formula f' thus obtained

$$g = \overline{x_k} + \overline{3} + f' + \overline{4}$$

g is then called a generalization of f.

Definition 26.[15] We understand by a "formula" a series of sets that occurs in the smallest set of series of sets that contains the elementary formulas and is closed with respect to negation, disjunction, and generalization.

Each formula is either an elementary formula, or it is a negation or generalization or disjunction of other such formulas, and it is in fact easy to convince oneself that only one of the four mentioned cases can enter in each case, and that in the case of a formula that is the disjunction of two others, also the members of the disjunction are uniquely determined. This follows from the way we use the bracket symbols. We call the number of bracket symbols divided by two the grade of a formula.

Definition 27.[16] We associate now to each

352R

formula its truth [value], by the use of induction after the grade.

1·) An elementary formula [written above: (formula of grade 0)] shall have truth value true when and only when (cf. Definition 21):

a. The class of level k is contained in one of level $k+1$.

b. The number designated by a number symbol (cf. Definition) is contained in a class of level 1 that stands in front of it.

c.) When it is identical with $\overline{5}$;

the truth value false in all other cases.

2·) A formula f of grade $k+1$ is either a disjunction of two formulas of lower grade, or a negation, or a generalization of a formula of lower grade. The truth value of f is determined in the usual way in the first two cases. In the second [third], f shall receive the truth value true if and only if all formulas from which it arises through generalization have the truth value true, in other cases the truth value false. By the remark made above, hereby every formula is assigned a truth value in a unique way.

One can make from each formula a form in which classes of corresponding types are substituted for free variables and truth values for propositional variables (i.e., the symbols 5, 6). Each proposition, especially,

[15] [There is no definition 25.]

[16] [This has been added later in the text that continues from the previous sentence.]

353L

is itself already a form. We define now:

A formula has the truth value true if and only if each form that arises through substitution has the truth value true.

We go now over into the definition of the concept provable.

Definition. x_3 is the result of detachment from y_3 and z_3 =

$z_3 = N(y_3)Ox_3$

Definition. x_3 is a ground axiom those given below

15 formulas that one obtains from the axioms given above on page when one replaces the letters by numbers in corresponding ways.

Definition. x_3 is k-fold type elevation of y_3

x_3 arises from y_3 when one replaces every variable of type m that occurs in it by one of type $m + k$ (the sign 1 must naturally not occur in y_3).[17]

$l(x_3) = l(y_3)$

$x_3(n) = x_3(m) \sim y_3(n) = y_3(m)$

If $x_3(n)$ not a variable, then $x_3(n) = y_3(n)$

If $x_3(n)$ propositional variable, then also $y_3(n)$ propositional variable

If $x_3(n)$ variable of type s, then $y_3(n)$ variable of type $s + k$

Definition. x_3 is an axiom shall mean

x_3 is a ground axiom or a type elevation of a ground axiom.

353R

Definition. x_3 is a subformula of y_3 [18]

$(En) \quad x_3(k) = y_3(k + n)$

$1 \leqq k \leqq l(x_3)$

[17] [This definition is put in brackets and a changed one as given here written on the next page.]

[18] [Added in margin: An elementary formula that is a subformula of x_3 is called an elementary component of x_3.]

Definition. We call a set to which belong all and only the subformulas of x_3 the subformula set of x_3.

Definition. Let S be a function always defined for all elementary components of x_3 such that $S(x_3)$ is [cancelled: a formula distinct from x_3.]

We define y_3 [cancelled: arises through substitution] $= S \mid$ (from x_3):

If there is a mapping F from the subformula set of x_3 to a set B of series of signs

354L

such that

1. For elementary components of e_i in x_3:

   $F(e_i) = S(e_i)$

2. F maintains the relations of negation, disjunction, and generalization.

   That is: From $m = N(k)$ must follow

   $F(m) = N[F(k)]$

   and from $m = u\, Gen\, F(k)$

   $F(m) = u\, Gen\, F(k)$

   and analogously for disjunction

3. $F(x_3) = y_3$

One sees easily that there is always one and just one y_3 of the required property.

Definition. Let x_3 be an elementary formula, v a variable that occurs in it, and A an arbitrary series of signs. Then we designate by [incomplete].

Since v stands always in the first or last place, we have

$$x_3 = \overline{v} + x_3' \qquad x_3' = x_3 - \overline{v}$$

or $$x_3 = x_3' + \overline{v}$$

354R

We designate now by "$x_3\ v|A$" the series of signs

$$y_4 = A \dotplus x_3'$$

or $\quad = x_3' \dotplus A$

Let x_3 be a formula wherein the free variable v occurs. We designate now by $x_3; v|A$ that [phrase seemingly incomplete]

Definition. $S|(x_3)$

Here $S(u_3) = u_3$ for all elementary components from x_3 that do not contain v.

$S(u_3) = u_3\ v|A$ for all elementary components from x_3 that contain v.

Let x_3 be a formula and v a variable that stands in the first place in each elementary component of x_3 in which it occurs. Let further A be an expression and w a free variable that occurs in A, of a level one less than that of v.

Definition. Then we designate by $x_3; v|A(v)$

$S|(x_3)$ where

355L

$S(u_3) = u_3$ for all elementary components from x_3 that do not contain v

$S(u_3) = A\, w|u_3 - \overline{v}$ for those elementary components from x_3 that contain v

Definition. y_3 is called a substitution result of x_3

when y_3 and x_3 are formulas and there is a formula z_3 distinct from x_3 such that there exist a propositional variable u, or variables v_k, v_{k-1} of types k and $k-1$ such that

$$y_3 = x_3; u|z_3$$

or $\quad y_3 = x_3; v_k|z_3(v_{k-1})$

Definition. We can now define the class of provable formulas as the smallest class of formulas that contains the axioms and is closed against the relations of detachment, generalization, and result of substitution.

355R

It is easy to convince oneself by complete induction of the correctness of the theorem:

Each provable formula is true.

For obviously the axioms are true formulas and this property is not destroyed by the rules of inference. This theorem can be proved, though, only with the help of the axiom of choice (the proposition that the formula of the axiom of choice is true means obviously the same as the axiom of choice).

Definition. The symbol $(a, \alpha_1 \alpha_2 ... \alpha_k)$ shall have a sense when and only when a is a formula with exactly k free variables. The α_i are allowed classes that coincide respectively with the types of the free variables of a, under the condition that the above symbol means those formulas that arise from a when the free variables therein are replaced by classes α_i (in the case of classes of type 0 [added above: i.e., numbers] by their symbols in accordance with definition), and the greater variable (for variables are numbers) by a class with a greater index.

Definition. A k-place relation between allowed classes of respective types $t_1 ... t_k$ shall be called finitely definable when there exists a formula a with exactly k free variables of the types $t_1 ... t_k$ such that

$$W(a; x_{t_1} ... x_{t_k}) \sim R(x_1 ... x_k)$$

in which x_i is a class of type t.

356L

It is easy to convince oneself that if two relations R_1 and R_2 are definable (through a_1 and a_2), then also $R_1 \vee R_2$ is definable through $a_1 \, O \, a_2$, and the same with the negations of R_1 and R_2 through $N(a_1)$ and $N(a_2)$, respectively. It holds further that when the n-place relation $R(x_1 ... x_n)$ is defined, then also the $n-1$-place relation $(x_1) R(x_1 ... x_n)$, and the relation is further definable that holds between a class of type $k+1$ and one of type k when the second is an element of the first one. From this it follows that every relation is definable that is built by the operations of $(\,)\ \vee\ ^{-}$ from the ε-relations (of different types). If one goes through the above definition, one recognises especially that the class of provable formulas is definable. Formulas are here conceived as relations between two natural numbers, so as classes of level 3 by a stipulation made above (cf.). One is convinced by complete induction, without further ado, of the theorem: Each provable formula is true.

We call the greatest number that occurs in a series of signs a the height of a ($h(a)$) and define an order relation [*vor*, before] for series of signs through the stipulation:

356R

Definition. x_3 *vor* y_3 when [?]

either $h(x_3) + l(x_3) < h(y_3) + l(y_3)$

or $h(x_3) + l(x_3) = h(y_3) + l(y_3)$

and $l(x_3) < l(y_3)$

or when $l(x_3) = l(y_3)$ $h(x_3) = h(y_3)$

and when $x_3(m) < y_3(m)$ for the smallest m

for which $x_3(m) \neq y_3(m)$

We define further a counting [Zählung] of all of the relation signs through the stipulation:

Definition. n *Zähl* $x_3 \underset{Df}{=}$

$n - 1$ is the cardinal number of the class of relation signs y_3 for which y_3 *vor* x_3.

It is easy to convince oneself that the relation n *Zähl* x_3 is definable.

Definition. Now a relation $R(ik)$ between natural numbers becomes as follows:

357L

$$R(nk) = \overline{Bew\{[(\imath x_3)nZx_3];n,0\}}$$

The relation $b = (a;ik)$ in which b denotes a proposition, a a relation sign, and i, k natural numbers, is obviously definable, therefore also the relation $R(ik)$, because it is built up from the concepts *Bew*, xZx_3, and $(a;i,k) = b$ with the help of the logical operations. So there is a relation sign a such that

$$W(a;ni) \sim \overline{Bew[\qquad\qquad]}$$

a occurs as a relation sign in the counting Z, say at the m-th place,

i.e., mZa

We form the following proposition $(a; m, 0)$ and claim that it is undecidable, for assuming $Bew(a; m, 0)$ we would also have

$$W(a; m, 0)$$

therefore $R(m, 0)$, i.e.,

357R

$\overline{Bew(Z(m); m, 0)}$ or

$\overline{Bew(a; m, 0)}$ i.e., we are at a contradiction.

From $BewN(a; m, 0)$ follows analogously

$$Bew(a; m, 0)$$

so again a contradiction, for when $(a; m, 0)$ as well as $N(a; m, 0)$ are both provable, both should be also true, impossible by the definition of the concept "true." So we have shown that the proposition

$$(a; m, 0)$$

is undecidable. One can give the formula a *in extenso*. For that, one needs to write down the definition of $R(ik)$ in the symbols of the *Principia Mathematica*. Even the number m can be determined by definition. So one can even write down $(a; m, 0)$ *in extenso*.

5. A proof in broad outline will be sketched

358L

In what follows, a proof in broad outline will be sketched by which the Peano axioms, together with the logic of the *Principia Mathematica* (with natural numbers as individuals), do not form a system definite with respect to decision, not even when the axiom of choice is included, i.e., that there exist unsolvable problems therein, and even of a relatively simple structure.

We replace the basic signs of the formal system **S** characterised {logical constants, variables of different types, successor, bracket symbols} by natural numbers in a one-to-one way, and correspondingly the formulas of system **S** through finite sequences of natural numbers. Thereby many metamathematical propositions *about* system **S** become (as they are indeed propositions about finite sequences of numbers) expressible *within* the system, something that is essential for the proof that follows.

Concepts used in what follows:

1. "Z-formula" = finite sequence of numbers that corresponds to a formula of system **S**
2. "Z-proposition" = "Z-formula" without free variables
3. "true Z-proposition" = "proposition" of the kind that the associated sentence of system **S** is true
4. "class sign" = "formula" with one free individual variable
5. $F(n)$ We think of the class signs as lexicographically ordered and denote the n-th by $F(n)$

How this ordering is taken makes no difference, just that one ordering is to be kept once and for all.

358R

6. *Neg* of a "formula" f $[N(f)]$ = the "formula" that arises from f through the setting ahead of the number that corresponds to the negation sign
7. "proved" is what a "formula" is called if it follows by the rules of inference from the axioms of system **S**
8. Let a stand for a class sign and n for a natural number. We denote by

[a; n] the "proposition" that arises from the class sign a when one substitutes therein the sign for number n in place of the free individual variable.

Sign for number n = the finite number sequence that corresponds to the following sign of system **S**:

$\underbrace{R'R' \dots R'0}_{n}$ (R = successor relation)

The most important concept for what follows is:

"Finitely definable" is what a class (of arbitrary type) K is called if there exists a class sign such that [incomplete sentence]

359L

1. Introduction (for easier expression of the theorem)
2. System of number theory (here also finite sets at hand)
3. Extension
4. One-to-one association between formulas and numbers. Metamathematical properties = properties of numbers
   occur in the system itself
5. Concept true
6. Concept finitely definable
7. Theorem expressed stated in different forms
8. Carrying through of the proof by the lemma
9. Proof of the lemma

359R

$F(n)$ the n-th class sign

$Klsz(n) \sim Form(n) \,\&\, (E!x)\{x\, Frva\, n \,\&\, x \leqq Höh(n)\}$

$F(n) = m$ the number-theoretic formula n gets expressed through m

$W\, Subst\, A\binom{11 \quad 12}{Z(x) \; Z(y)} \sim x\, Bew\, y$

$(\overline{Ex})Bew[n; Z(n)]$

$$W[q; x, n] \sim x\, Bew\, F[n, Z(n)]$$

$$p = NE, q$$

$$W[p; n] \sim (\overline{Ex}) Bew\, F[n; Z(n)]$$

$$F[p; Z(p)] \quad F[N(p; Z(p)]$$

6. We produce an undecidable proposition in the *Principia*

360L

The development of mathematics in the direction of greater exactness has led, as is well known, to wide areas of it being formalized (as intended, the whole of mathematics).[1] The most comprehensive formal systems put up at present are the *Principia Mathematica* on the one hand, and the Zermelo-Fraenkelian axiom system of set theory (as developed further by J. von Neumann). Both of these systems are so wide that all the proof methods used as of today in mathematics, formalized in them, are led back to a few axioms and rules of inference. Therefore, the conjecture lies close at hand that these axioms and rules of inference are really sufficient to carry through each proof thinkable in general. In what will be presented, it is shown that this is not the case, but that there are instead in both of the systems referred to[2] problems[3] that

360R

cannot be decided with the help of the axioms and rules of inference put up. It is a situation that lies in no way in, say, the special nature of the systems put up so far, but holds in general for any formal system in which number-theoretic problems can be expressed and in which axioms are at hand only in a finite number or are obtained through substitution in a finite number of steps. It has to be required further that no false number-theoretic propositions are provable in a system.

The formulation just given is still vague and will be made precise in what follows. Nevertheless, we want to sketch now the course of the proof in rough outline, for an orientation for the reader, naturally without raising any pretence to exactness. To fix ideas, we consider the system of the *Principia*. The propositions of this system are, externally considered, finite series

---

[1] [Cancelled but intermittently underlined for emphasis: in a way in which proving is carried through by a few mechanical rules.]

[2] Here we assume in the *Principia*, in particular, the axiom of infinity in the form, there exist exactly denumerably many individuals, naturally the axiom of reducibility, and, if one wishes, also the axiom of choice.

[3] Indeed, there exist in particular problems in which no other concepts than + (sum) and $\times$ (product) (both applied to natural numbers) occur, further the logical concepts $=$ $(\ )$ $\overline{\ }$ $\vee$, in which $(\ )$, for all, is allowed to apply only to natural numbers.

of basic signs, and it is easy to make precise when a finite series of basic signs is a meaningful proposition

361L

and when not. [Added: We think here and in what follows of each proposition as having been written down without abbreviations.] The proofs, in turn, are finite sequences of propositions, with certain properties that can be given precisely. (Each proposition in a proof must be either an axiom or arise from some of the ones before through the application of the rules of inference.)

It is obviously indifferent for metamathematical considerations what one takes as basic signs. We can take, in particular, natural numbers as basic signs, and a proposition is then, correspondingly, a finite sequence of natural numbers (with certain properties that can be given precisely) and a proof a finite sequence of finite sequences of natural numbers. Thereby all metamathematical concepts and propositions become concepts and propositions about natural numbers or finite sequences of natural numbers etc. Now, because natural numbers occur within the system of the *PM* (this meant in a contentful way), a greater part of metamathematics becomes thereby expressible within the system itself (this meant contentfully). One can show, especially, that the concepts of formula, proof, provability [written above: *Bew*] can be defined in the *PM* itself, i.e., one can give, for example, a formula $F(x)$ of the *Principia* with one free variable, such that $\Gamma(x)$ states, contentfully interpreted: x is a provable formula. [Added: It is very easy (just somewhat long-winded) to actually write down this formula.]

361R

We produce now an undecidable formula of the *Principia* in the following way:

We shall call a formula of the *Principia* with one free variable of the type of class of classes [added: this is the type of the natural numbers in the *Principia*] a class sign. We think of the class signs as lexicographically ordered and designate the n-th by $F(n)$ [added remark: There are, admittedly, infinitely may letters, namely the natural numbers, but that is obviously no obstacle.] and note that the ordering relation P lets itself be defined within the *Principia*. We designate by $[P(n);k]$ the formula that arises from the class sign $P(n)$ when one replaces the free variable by the sign for the natural number k. Even the triple relation $z = [x;y]$ turns out to be definable within the *PM*. We define now a class K of natural numbers by the condition:

$$K(n) \underset{Df}{=} \overline{Bew[P(n);n]}$$

All the concepts that appear on the right, namely Bew, P, $[x;y]$, are definable in the *PM*, therefore also the concept K composed of them, i.e., there exists a class sign R[4]

362L[5]

such that formula $[R;n]$ states, meant contentfully, that $K(n)$ holds.

As a class sign, R is identical to some $P(x)$, i.e.,

$$R = P(q) \quad \text{for a determinate } q$$

We show now the *following theorem:* The proposition $[P(q);q]$[6] of the *Principia* is not decidable from the axioms.

For were the proposition $[P(q);q]$ provable, it would be, contentfully interpreted, correct. Then $K(p)$, i.e.,

$$\overline{Bew[P(q);q]}$$

would hold, in contradiction with the assumption. If instead the negation of $[P(q);q]$ were provable, then $\overline{K(n)}$, therefore $Bew[P(q);q]$ would hold. Then $[P(q);q]$ together with its negation would be provable which is again an impossibility.[7]

The analogy of this proof with the antinomy of Richard

362R

hits the eye. There is even a close resemblance with the Cretan inference, for the undecidable proposition $[F(p);p]$ states by the above that $[F(p);p]$

[4] Again, there is not the least difficulty in actually writing down the formula R.

[5] [The second half of this page is quite faint, but the text can still be read with the help of the typewritten version that is very similar.]

[6] As soon as one has actually determined R, then q and thereby the undecidable proposition can be effectively written down.

[7] The basis of the treatment as a whole lies in the possibility of a contentful interpretation of the propositions of the *Principia* (a specific exact consideration would be required for this). The freedom from contradiction of the system results of course all by itself from this supposition. [Remark added at bottom of the facing page.]

is not provable.[8] So we have a proposition in front of us that claims its own unprovability.[9]

The whole train of thought can be carried over – –[10]

What is the logical structure of the undecidable proposition? It claims, by the above, the inexistence of a proof for $[P(q;q)]$, i.e., the inexistence of a finite sequence of finite sequences of natural numbers with a property F, to be a proof for $[P(q;q)]$. One can map the finite sequence of sequences of natural numbers in a unique way to the natural numbers, and hereby the undecidable proposition assumes the form

$\overline{En)}F(n)$[11]

in which F denotes a property of natural numbers.

363L [The first half of this page is cancelled, the rest very weak]

One can establish through a precise analysis (to be conducted in what follows) the following concerning the property F:

1. It is definite with respect to decision, i.e., one can decide for every natural number whether it applies or not (and even inside the system of the *Principia*).

363R

For one can, on the one hand, decide of each formula whether it is an axiom or not, on the other hand whether it follows from given other formulas by the rules of inference. Therefore it is even decidable for each sequence of formulas whether it is a proof and this extends to the numbers assigned to sequences of formulas. [Added in margin: The definiteness with respect to decision of F gets especially expressed through the possibility to define F recursively.[12]]

---

8 [The typewritten manuscript has $[R(q);q]$ instead of $[F(p);p]$.]

9 It seems at a first look as if such a proposition had to be either nonsense, because the object it talks about is constructed only through the proposition itself. The situation is, instead, rather the following: The proposition $[F(p);p]$ states, initially, that a specific, precisely given formula is unprovable. It turns out later, and in a certain way by chance, that this formula is precisely the one in which it itself gets expressed.

10 [The typewritten manuscript continues from this point on in a different way.]

11 $F(n)$ means here: n is associated to a proof of $[F(p);p]$.

12 Definable by recursion means always: definable by the simplest form of recursion (without function parameters).

2. The property F can be defined with just the concepts + (addition), × (multiplication) (both concerning natural numbers), = identity, ( ) for all (concerning natural numbers), ∨ (or), $^{-}$ negation. I.e., there are already in the arithmetic[13] of natural numbers built merely upon addition and multiplication problems that are not solvable by the means of the axioms of the *Principia* (and not even those of set theory).

It will be shown below that a problem of the form $(En)F(n)$, in which F is a property defined by simple recursion, is always equivalent to the question whether a certain formula $G(\varphi, \psi)$ of the narrower functional calculus[14] (determined through F) is generally valid or not. This equivalence can be also proved within the *Principia*.

364L

One concludes therefore [added above: theorem from 1] that there are in the *PM* formulas within the narrower functional calculus for which neither general validity nor the existence of a counterexample is provable.

In the above proof of the undecidability of $[P(q);q]$, we have made use of the concept of contentful correctness of formulas of the *Principia*, and especially of the theorem that every provable formula is contentfully correct. Because of the problematic nature of these things (one can think of, say, the axiom of choice), it would be advisable to avoid the concept of contentful correctness as far as possible and [?] to replace where possible the condition of contentful correctness of provable propositions by the freedom from contradiction of the system. This succeeds (at least in part) in the following way: We assume the concept of contentful correctness as given only for formulas of the form $F(n)$, where F is a recursively defined property and n denotes a specific natural number. This concept is, admittedly, completely unproblematic for such formulas and beyond that

364R

exactly defined. Further, we shall call a system ω-consistent when the following holds: For no recursive property G is

[13] Observe in particular that there occur *no* functions, just variables for natural numbers, in the propositions within this domain.

[14] φ, ψ shall be the function variables that occur in G (all of them are, by the definition of the narrower functional calculus, free variables, cf. Hilbert-Ackermann).

$(Ex)\overline{G(x)}$ together with all the formulas

$G(1)\ G(2)\ .\ .\ G(n)$ *ad inf.* provable

Then the following holds:

If $(Ex)G(x)$ ($G(x)$ recursive) is provable in an ω-consistent system in which the proofs of finite number theory can be carried through, then there exists an n for which $G(n)$ is contentfully correct. For in other case, $\overline{G(n)}$ would be correct for all n, therefore (because of the recursive definability of G), even provable. So, $(Ex)G(x)$ would be provable together with all of the $\overline{G(n)}$, against the assumption.[15]

Further, for each system free from contradiction holds already:[16]

In case $(x)G(x)$ is provable (G recursive), $G(n)$ is contentfully correct for all n, for were for one, say for m, $\overline{G(m)}$ correct, it would also be provable

365L

in contradiction with $(x)G(x)$.

One can, then, make for ω-consistent (and sufficiently wide) systems the inference from the provability of propositions of the form of $(x)G(x)$, $(Ex)G(x)$ (G recursive) to their contentful correctness, and precisely this is used in the above proof. Therefore it holds word for word also for all ω-consistent extensions of the *PM*, as soon as the class of axioms to be added is recursively definable (especially with finitely many ones).

As concerns the concept of ω-consistency, let the remark be added that there are extensions of the *PM* that are, indeed, free from contradiction, but they are not ω-consistent. One obtains such in the following way. One sees easily that of the two propositions $(En)F(n)$, $\overline{(En)F(n)}$, neither of which is provable,[17] the first is false, the second correct. For $(En)F(n)$ states that there exists a proof for the formula $[P(q);q]$ and this was the one shown to be undecidable.

365R

Then, for each n, $\overline{F(n)}$ is correct, therefore (by the recursive definability of

[15] [The negation lines in the $\overline{G(n)}$ are not visible.]

[16] It is assumed that finitary number-theoretic proofs can be carried through in this system.

[17] F is the property defined on page .

F) also provable. So, if one adds $(En)F(n)$ to the *Principia*, one obtains a system that is free from contradiction but not ω-consistent.

The fact just used, namely that problems undecidable in the system of the *PM* still can be decided in a metamathematical way, is even in itself of interest. An investigation of which means not available in the *PM* make possible a metamathematical proof of $\overline{(En)F(n)}$ leads to the result that the one and only such means that comes into consideration is the freedom from contradiction of the *PM*, as required in the metamathematics. I.e., if one could prove the freedom from contradiction of the *PM* within the *PM*, then one could even prove $\overline{(En)F(n)}$, which is in contradiction with the undecidability of this problem. From this follows the strange result that one cannot carry through a proof of freedom from contradiction for the *PM* even with all the logical means contained in the *PM*. It needs hardly be mentioned that all considerations carried through so far

366L

can be extended to arbitrary formal systems as soon as number theory is contained in them[18] and the axiom classes definable through ordinary recursion. There are in each such system, as soon as it is ω-consistent, unsolvable number-theoretic problems [cancelled: and there exists for no such system a proof of freedom from contradiction that uses only means contained in the system.]

[18] I.e., number-theoretic propositions (in a sense to be made precise below) must be expressible and number-theoretic proofs formally executable.

7. The development of mathematics in the direction of greater exactness

249R

The work undecidability, draft

1.

The development of mathematics in the direction of greater exactness has led, as is well known, to wide areas of it being formalized, in a manner in which proofs can be carried through by a few mechanical rules. The most comprehensive formal systems put up at present are the system of the *Principia Mathematica* (P) on the one hand, and the axiom system of set theory of Zermelo Fraenkel [cancelled: (M)] (developed further by *v. Neumann*). Both of these systems are so wide that all the proof methods used in mathematics today are formalized in them, i.e., led back to a few axioms and rules of inference. Therefore, the conjecture lies close at hand that these axioms and rules of inference are sufficient to carry through each thinkable proof in general. It will be shown in what follows that this is not the case but that there exist (in both of the systems put forward)[1] even arithmetic problems from the theory of ordinary numbers[2] that cannot be decided with the help of the axioms and rules of inference put up. It is a situation that lies in no way in, say, the special nature of the systems (P, M) put up, but

250L

holds instead for a very wide class of formal[3] systems to which belong especially all those that arise from M and P through the addition of finitely

[1] Here, we assume especially as axiom in P the axiom of reducibility. [The margin has very faintly: axiom of infinity, there exists exactly denumerably many]

[2] I.e., more precisely, there exist undecidable propositions in which there occur no concepts beyond the logical constants $^{-}$, $\vee$ ( ) $=$, except + (add) $\times$ (mult), both in relation to natural numbers, and in which even the prefix ( ) relates only to natural numbers. (In such propositions, there can thus occur only numerical variables, but never function variables whether free or bound.)

[3] To this class belong for example all formal systems the axioms of which arise through substitution of arbitrary formulas in finitely many schemes (as rules of inference the usual ones, say those assumed in the *PM*).

many axioms[4], with the condition that no false propositions of the kind given in[2)] become provable through the newly added axioms.

We shall sketch to begin with, before we go into the details, the main idea of the proof, naturally with no pretence to exactness. The formulas of a formal system (we delimit ourselves here to the system P of the *Principia*) are, externally considered, finite series of basic signs (variables and logical constants), and it is easy to make it precise *which* series of basic signs are meaningful formulas and which not.[5] Proofs are analogously, externally considered, nothing but finite series of propositions (with specific properties). It is, for metamathematical considerations, obviously indifferent what objects one takes as basic signs. We decide to use natural numbers as such signs.[6] A formula is then, correspondingly, a finite sequence of natural numbers[7] and a proof figure a finite sequence of finite sequences of natural numbers. The metamathematical concepts (propositions) become hereby concepts (propositions) about natural numbers (and sequences of such) and therefore (at least in part) expressible within the system P (this meant in a contentful way). One can show, especially,

250R

that the concepts "formula," "proof figure," "provable formula" are definable within the system P, i.e., one can, for example, give a formula $F(x)$ of one free variable of the *PM* such that $F(x)$ states, interpreted contentfully: x is a provable formula (that we abbreviate with the designation $Bew(x)$).[8] We produce now an undecidable proposition of the system P (i.e., a proposition A for which neither A nor $\overline{A}$ is provable) in the following way:

We shall call a formula from P with one free variable of the type of the natural numbers (class of classes)[9] a class sign. We think of the class signs as lexicographically ordered and designate the n-th by $R(n)$ and note that the concept of "class sign" as well as the ordering relation R let themselves

[4] In this, only those axioms are counted as distinct in P that do not come out one from the other by a simple change of type.

[5] We mean here and in what follows by "formula of the system P" only a proposition written without abbreviations (i.e., without the use of definitions).

[6] I.e., we map the basic signs in a one-to-one way on the natural numbers.

[7] In an abstract sense, i.e., as functions over segments of the natural number sequence of natural numbers. [Belegungen eines Abschnittes der Zahlenreihe mit natürliche Zahlen]

[8] It would be very easy (just a bit long-winded) to actually write down this formula.

[9] That is the type of the natural numbers in P.

be defined within P. We designate by $[P(n), k]$ the sentence-formula that arises from the class sign $P(n)$ through the replacement of the free variable by the sign for the natural number k. Even the relation $z = [x, y]$ [10] turns out to be definable within P. We define now a class K of natural numbers as follows:

251L

$$K(n) \underset{Df}{=} \overline{Bew}[R(n), n] \tag{1}$$

All of the concepts that occur in the *Definiens* are definable in P, and therefore also the concept K composed of them., i.e., there is a class sign S [11] such that formula $[S; n]$ states, meant in a contentful way, that the natural number n belongs to K. As a class sign, S is identical to a determinate $R(x)$, i.e., we have

$$S = R(q) \tag{2}$$

Here q is a determinate natural number. We show now that the proposition (more precisely, proposition-formula) $[R(q), q]$ [12] is undecidable in P. For if it is assumed that the sentence $[R(q), q]$ is provable, then it would even be correct contentfully interpreted, i.e., by the above, q would belong to K, i.e., by (1), $\overline{Bew}[R(q), q]$ would hold in contradiction with the assumption.

251R

If instead the negation of $[R(q), q]$ were provable, then $\overline{K(q)}$, i.e., $Bew[R(q), q]$ would hold. $[R(q), q]$ together with its negation would be provable which is again impossible.

The analogy of this inference with the antinomy of Richard hits the eye. There is even a close relation with the "liar" [changed from: Cretan inference], for the undecidable proposition $[R(q), q]$ states that q belongs to $K(n)$, i.e., by (1), that $[R(q), q]$ is not provable. So we have a proposition in front of us that claims its own unprovability.[13]

---

[10] In case x is not a class sign or k no natural number, one means by $[x, y]$ the empty sequence of numbers, say.

[11] There is, again, not the least difficulty actually to write down the formula S.

[12] As soon as S has been actually determined, even q lets itself obviously be determined and thereby the undecidable proposition effectively written down.

[13] Such a proposition has, contrary to appearance, nothing circular about it, for it claims in the first place the undecidability of quite a specific formula (namely the q-th in the lexi-

The proof method just presented can be evidently applied to every formal system that:

1. In terms of content, provides sufficiently in the form of means of expression, so that the concepts used in the above considerations (especially the concept provable formula) can be defined and in which

2. every provable formula is even contentfully correct.

These conditions are satisfied, especially, by the Zermelo Fraenkel axiom system of set theory.

The exact carrying through of the above idea [changed from: proof] that is to follow has, actually, as its main task the replacement of the concept "definable in the system" that is too vague by a precise one, to eliminate the suspicious concept of contentful correctness (think just of the axiom of choice, say) entirely from consideration.[14]

252L

2

We go now into the exact carrying through of the proof sketched above and give, to start with, a precise description of the formal system P for which we want to prove the existence of undecidable propositions. P is essentially the system one obtains if one builds, upon the Peano axioms,[15] the logic of the *Principia Mathematica* (numbers as individuals, successor relation as an undefined basic concept).

The basic signs of the system are as follows:

"$\sim$" (not), "$\vee$" (or), "Π" (for all, with the usage $x\Pi F(x)$)

"0" (zero) "f" (successor of) "(" ")" (bracket symbols)

Further:

variables of type 1 (for individuals, i.e., natural numbers) "x_1" "y_1"

cographical ordering). It turns out only afterwards (by chance as it were) that this formula is just the one in which the formula itself gets expressed.

[14] [Added above: to avoid completely the contentful interpretation of the formulas of the system considered]

[15] The addition of the axioms of Peano to the system is done just on technical grounds. [Added remark: From a formal point of view, Π as well as all the other opportune changes in system *PM* take place merely on technical founds.]

"z_1"...

variables of type 2 (for classes of natural numbers) "x_2" "y_2" "z_2"...

variables of type 2 (for classes of classes of natural numbers) "x_3" "y_3" "z_3"...

etc *ad inf*[16]

Variables for two-place functions (relations) are superfluous as basic signs, because one can define relations as classes of ordered pairs, and ordered pairs in turn as classes of classes, for example

252R

the ordered pair (a, b) by $[(a), (a, b)]$[17] in which (a, b) and (a) denote the classes the only the elements of which are a, b (and a, respectively).

We mean by a *sign of the first type* a combination of signs of the form "a" "fa" "ffa" etc

Here "a" is either "0" or a variable of type 1 (in the first case, the sign becomes a "*number sign*").

For $n > 1$, we mean by a *sign of the n-th type* the same as a *variable of "n-th type."*

We call combinations of signs of the form $a(b)$, in which b is a sign of type n and a a sign of type $n+1$, "*elementary formulas.*" We define the class of "*formulas*" as the smallest class of combinations of signs to which belong the elementary formulas and to which belong, with a and b, always also $(a) \vee (b)$ $\sim (a)$ $v\Pi(a)$ (here v denotes an arbitrary variable[18]). We call $(a) \vee (b)$ the *disjunction* of a and b, $\sim (a)$ the *negation* of a, and $v\Pi(a)$ a *generalization* of a.

A formula in which there occur no free variables ("free variables" defined in the usual way) is called a "*sentence sign.*" We call a formula with exactly n free individual variables, $n \geqq 1$, (and with no further free variables) an *n-place relation sign*.

---

[16] It is required that denumerably many signs are available for each type of variable.

[17] Even inhomogeneous relations can be defined in this way, for example, a relation between individuals and classes as a class with elements of the form $((x_1), x_2)$.

[18] So $v\Pi(a)$ is a formula even in the case that v does not occur or does not occur free in a (it then means naturally the same as a).

253L

We mean by $Sb(a\,{v \atop b})$ (in which a denotes a formula, v a variable, b a sign of the same type as v) the formula that arises from a when v is replaced in it (everywhere where it is free) by b.[19]

We say that a formula a is a *type elevation* of another b if a arises from b through the elevation by the same number of all the free variables that occur in b.

The following formulas are called "axioms" (they are written down with the help of the abbreviations $\rightarrow\ \&\ \equiv\ (Ex)\ =$[20]):

I[21] 1.) $\sim(fx_1 = 0)$

2.) $fx_1 = fy_1 \rightarrow x_1 = y_1$

3.) $x_2(0)\,\&\,x_1\Pi(x_2(x_1) \rightarrow x_2(fx_1)) \rightarrow x_1\Pi(x_2(x_1))$

II Each formula that arises from the following schemes through the substitution of arbitrary formulas for p, q, r:

1 $p \vee p \rightarrow p$

2 $p \rightarrow p \vee q$

3 $p \vee q \rightarrow q \vee p$

4 $(p \rightarrow q) \rightarrow (r \vee p \rightarrow r \vee q)$

253R

III Each formula that arises from the two schemes

1.) $x\Pi A \rightarrow Sb(A\,{x \atop b})$

2.) $x\Pi(B \vee A) \rightarrow B \vee x\Pi A$

through making the following substitutions for A, B, x:

for A an arbitrary formula

for x an arbitrary variable

---

[19] In case v is not free in a, $Sb(a\,{v \atop b}) = a$ shall be the case.

[20] $v = w$ is, as in the *Princ. Math.*, defined by [the typewritten version, footnote 21, suggests the completion: $x_2\Pi(x_2(v) \equiv x_2(w))$].

[21] [Gödel gives just the numbers for axioms in I and II. I have added the axioms from the typewritten manuscript. The logical notation in the typewritten version is somewhat different from the one here.]

for B a formula in which x does not occur free

for b a sign of the same type[22] as x, on the condition that b contain no variable (or is such) that is bound in A in one of the places in which x is free.[23]

IV Each formula that arises from the following scheme through the substitution of arbitrary variables of the types $n, n+1$ for v, u and a formula for A that does not contain u free:

$$(Eu)(v)[u(v) \sim A]^{24}$$

V Each axiom that arises from the following one through type elevation:[25]

$$(x_1)x_2(x_1) \equiv y_2(x_1) \rightarrow x_1 = x_2$$

This axiom states that a class is completely determined by its members.

254L

A formula c is called an *immediate consequence* of a and b (or of a) if a is the formula $b \rightarrow c$ (or if $c = x\Pi b$, respectively, where x is an arbitrary variable). The class of *provable formulas* is defined as the smallest class of formulas that contains the axioms and is closed with respect to the relation of "*immediate consequence*."[26]

We associate next natural numbers to the basic signs, in the following one-to-one way:

"0" $\rightarrow$ 1 "f" 2 (3) 4 $\sim$ 5 $\vee$ 6 Π 7

Further, for the variables of type n numbers of the form p^n in which p denotes a prime number > 7. Hereby there corresponds to each finite series of basic signs (therefore also each to formula) a finite series of natural numbers in a one-to-one way. We map now the finite series of natural numbers (again in a one-to-one way) on the natural numbers through the stipulation:

22 So b is either a variable or a sign of the form $ff\ldots fx_0$ in which x_0 is either 0 or a number variable.

23 [The previous footnote is repeated here.]

24 This axiom means the same as the axiom of reducibility (the separation axiom in set theory).

25 [The conclusion in Gödel's axiom should be $x_2 = y_2$ as in the typewritten manuscript.]

26 The rule of substitution becomes superfluous through the carrying out of all possible substitutions already in the axioms.

254R

To the series $e_1 e_2 \dots e_n$ shall correspond the number p:

$2^{e_1} 3^{e_2} \dots p_n{}^{e_n}$ in which p_n is the n-th prime number.

Hereby a natural number is associated in a one-to-one way, not just to each basic sign but also to each finite series of basic signs. We designate the number associated to the basic sign (or the finite series of basic signs) a by $\varphi(a)$.

Next, let there be given whatever class or relation $R(a_1 \dots a_n)$ $(n \geqq 1)$ between basic signs or series thereof. We associate to it that class (relation) $R'(x_1 \dots x_n)$ between natural numbers that obtains between $x_1 \dots x_n$ when and only when there exist $a_1 \dots a_n$ such that $x_i = \varphi(a_i)$ $(i = 1, 2, \dots n)$ and $R(a_1 \dots a_n)$.

We designate the classes (relations) between natural numbers that are associated in this way to the metamathematical concepts so far defined, for example "variable" "elementary formula" "formula" "sentence-sign" "negation" "axiom" "immediate consequence" "provable formula" "$x = Sb(a\,{}^{v}_{b})$," etc, by the same words but in cursive writing. The existence of formally unsolvable problems becomes thereby expressed within the system P as follows: There exists a *sentence formula a* such that neither *a* nor the *negation* of *a* are *provable* formulas.

We put next up the following definition: A number-theoretic function[27] $\varphi(x_1 \dots x_n)$ (with arbitrarily many independent variables)

255L

is said to be recursively defined from the number-theoretic functions

$\psi(x_1 \dots x_{n-1})$ and $\mu(x_1\, x_2)$

if the following holds for all $x_2 \dots x_n\, k$:

$$\begin{aligned} \varphi(0\, x_2 \dots x_n) &= \psi(x_2 \dots x_n) \\ \varphi(k{+}1\, x_2 \dots x_n) &= \mu(k, \varphi(k\, x_2 \dots x_n)) \end{aligned} \tag{1}$$

Further, a number-theoretic function is said to be recursive if there exists a finite series of functions $\varphi_1 \dots \varphi_n$ that ends with φ and has the property that each of the functions φ_k of the series is either recursively defined from

[27] I.e., its domain of definition is the class of non-negative entire numbers and its range of values a (proper or improper) subclass thereof.

two preceding ones, or arises from whichever of the preceding through substitution,[28] or finally is a constant or identical to the successor function (x_1+1) (or x_2+1, etc). The length of the shortest series of φ_i that belongs to a recursive function is called its level [Stufe].

A relation (class) between natural numbers $R(x_1 \dots x_n)$ is recursive if there is a recursive function $\varphi(x_1 \dots x_n)$ such that

$$R(x_1 \dots x_n) \equiv \varphi(x_1 \dots x_n) = 0^{29}$$

The following theorems hold:

I Each (function) relation that arises from recursive functions (relations) through substitution of recursive functions in the place of variables is recursive. The same for functions that arise from recursive functions by recursive definition according to scheme (1).

II If R and G are recursive relations, then also $\sim R \;\; R \vee G \;\; R \,\&\, G$.

255R

III If the functions $\varphi(\mathfrak{x})$ and $\psi(\mathfrak{y})$ are recursive, then also the relation

$$R(\mathfrak{x}\mathfrak{y}) \;=\; \varphi(\mathfrak{x}) = \psi(\mathfrak{y})^{30}$$

IV If the function $\varphi(x\mathfrak{x})$ [31] and the relation $R(x\,\mathfrak{y})$ are recursive, then also the relations

$$S(\mathfrak{x}, \mathfrak{y}) \equiv (Ex)[x \leqq \varphi(\mathfrak{x}) \,\&\, R(x\,\mathfrak{y})]$$

$$T(\mathfrak{x}, \mathfrak{y}) \equiv (x)[x \leqq \varphi(\mathfrak{x}) \rightarrow R(x\,\mathfrak{y})]$$

as well as the function

$$\varepsilon_x\{x \leqq \varphi(x\mathfrak{x}) \,\&\, R(x\,\mathfrak{y})\} = \psi(\mathfrak{x}\mathfrak{y})$$

in which $\varepsilon_x f(x)$ means: the smallest x for which $f(x)$ holds or 0 (in case there is no such x).

Theorem I follows at once from the definition.

[28] More precisely, through substitution of some of the preceding functions in the argument places of others that precede, so for example:
$\varphi_k(x_1 x_2) = \varphi_l\{\varphi_p(x_1), \varphi_q(x_1 x_2) x_1\}$ $l, p, q < k$.

[29] Recursive relations have obviously the property that one can decide for each specific n-tuple of numbers whether $R(x_1 \dots x_n)$ holds.

[30] We use German letters $\mathfrak{x}\,\mathfrak{y}$ as abbreviatory expressions for series of arbitrarily many variables $(x_1 x_2 \dots x_n)$.

[31] [IV has twice $\varphi(\mathfrak{x})$ with x cancelled, twice $\varphi(x\mathfrak{x})$, the former as in TM.]

Theorems II and III depend, as one easily convinces oneself, on the recursiveness of the functions

$$\delta(x) \quad \mu(xy) \quad \nu(xy)$$

that correspond to the logical concepts $\sim, \vee, =$, namely

$$\delta(0) = 1 \quad \delta(x) = 0 \text{ for } x \neq 0$$

$$\mu(0\,x) = \mu(x\,0) = 0 \quad \mu(x\,y) = 1 \text{ when } x\,y \text{ both} \neq 0$$

$$\nu(x\,y) = 0 \text{ when } x = y \quad \nu(x\,y) = 1 \text{ when } x \neq y$$

256L

The proof of theorem IV is in brevity the following:

There is by assumption a recursive ϱ such that $\varrho(x\,\mathfrak{y}) = 0 \equiv R(x\,\mathfrak{y})$. We define now by recursion scheme (1) a function $\chi(x\,y)$ in the following way:

$$\chi(0\,\mathfrak{y}) = 0$$

$$\chi(n+1, \mathfrak{y}) = (n+1)\,.\,a + \chi(n\,\mathfrak{y})\,.\,\delta(a)$$ [32]

Here $a = \delta\varrho(n+1, \mathfrak{y})\delta\delta\varrho(0\,\mathfrak{y})\,\delta\chi(n\,\mathfrak{y})$

$\chi(n{+}1\,\mathfrak{y})$ is therefore either $= n+1$ (when $a = 0$) or $= \chi(n\,\mathfrak{y})$ (when $a = 1$).[33] The first case occurs clearly if and only if all of the factors of a are 1, i.e., when the following holds:

$$\sim R(0\,\mathfrak{y})\ \&\ R(n{+}1\,\mathfrak{y})\ \&\ (\chi(n\,\mathfrak{y}) = 0)$$

From this it follows that the function $\chi(n\,\mathfrak{y})$ (considered as a function of n) remains 0 until the smallest value

256R

for which $R(n\,\mathfrak{y})$ holds, and is from there on equal to this value (in case that $R(0\,\mathfrak{y})$ already holds, we have correspondingly $\chi(n\,\mathfrak{y})$ constant = 0).

By this we have

$$\psi(\mathfrak{x}\,\mathfrak{y}) = \varepsilon_x\{x \leqq \varphi(x\mathfrak{x})\ \&\ R(x\,\mathfrak{y})\} = \chi(\varphi(x\,\mathfrak{x}, \mathfrak{y})$$

[32] We assume as known that the functions $a+b$ (addition) $a\,.\,b$ (product of two numbers) are recursive.

[33] a cannot apparently have other values.

$$S(\mathfrak{x}\,\mathfrak{y}) \equiv R(\psi(\mathfrak{x}\,\mathfrak{y}),\mathfrak{y})$$

The relation T can be reduced through negation back to a case that is analogous to S, by which theorem IV is proved.

The functions $a+b$, $a\,.\,b$, a^b, further the relations $a<b$, $a=b$ are, as one is easily convinced, recursive, and starting with these concepts, we define now a series of functions (relations) (1–) each of which is defined from the preceding ones by the procedures mentioned in theorems I–IV. Here, several of the steps of definition allowed by theorems I–IV are usually combined into one. Each of the functions (relations) 1– (among them occur for example the concepts "formula" "axiom" "immediate consequence") is therefore recursive.

1. $x/y \underset{Df}{=} (Ez)\{z \leqq x \,\&\, x = yz\}$

   x is divisible by y

257L

2. $Prim(x) \underset{Df}{=} \overline{(Ez)}\{z \leqq x \,\&\, z \neq 1 \,\&\, z \neq x \,\&\, x/z\} \,\&\, x > 1$

   x is a prime number

3. $0\,Pr\,x = 0$

   $(n+1)Pr\,x = \varepsilon_y\{y \leqq x \,\&\, Prim(y) \,\&\, x/y \,\&\, y > n\,Pr\,x\}$

   $n\,Pr\,x$ is the n-th (in order of value) prime number contained in x

4. $0! = 1$

   $(n+1)! = n! \cdot (n+1)$

5. $Pr(0) = 1$

   $Pr(n+1) = \varepsilon_y[y \leqq [Pr(n)]! + 1 \,\&\, Prim(y) \,\&\, y > Pr(n)]$

   $Pr(n)$ is the n-th prime number (in order of value)

6. $n\,Gl\,x = \varepsilon_y[y \leqq x \,\&\, x/(n\,Pr\,x)^y \,\&\, x/(n\,Pr\,x)^{y+1}]$ [34]

   the n-th member of the number series that corresponds to x (in accordance with the stipulation set on page)

[34] [The last conjunct is negated in TM.]

257R

7. $l(y) = \varepsilon_x\{x \leqq y \,\&\, x\,Pr\,y > 0 \,\&\, (x+1)\,Pr\,y = 0\}$

   the length of the number series that corresponds to y

8. $x \mathbin{+\!\!\!+} y = \varepsilon_z\{z \leqq [Pr(lx + ly)]^{x+y} \,\&\, l(z) = l(x) + l(y)$

   $\&\, (n)[n \leqq lx \rightarrow n\,Gl\,z = n\,Gl\,x] \,\&$

   $\&\, (n)[0 < n \leqq l(y) \rightarrow (n + lx)\,Gl\,z = n\,Gl\,y]\}$[35]

   $x \mathbin{+\!\!\!+} y$ corresponds to the operation of "adjoining one to another" of two finite series of numbers

9. $R(x) = 2^x$

   $R(x)$ corresponds to the number series that consists of just the number x

10. $E(x) = R(3) \mathbin{+\!\!\!+} x \mathbin{+\!\!\!+} R(4)$

    this corresponds to the operation of *bracketing* (3 and 4 represent the signs (and))

11. $n\,Var\,x = (Ez)\{z \leqq x \,\&\, Prim(z) \,\&\, x = z^n \,\&\, x > 7\}$

    x is a *variable* of type n

12. $Var(x) = (En)\{n \leqq x \,\&\, n\,Var\,x\}$

    x is a *variable*

258L

13. $Neg(x) = R(5) \mathbin{+\!\!\!+} E(x)$

    negation of x

14. $x\,Od\,y = E(x) \mathbin{+\!\!\!+} R(6) \mathbin{+\!\!\!+} E(y)$

    disjunction of x and y

15. $x\,Gen\,y = R(x) \mathbin{+\!\!\!+} R(7) \mathbin{+\!\!\!+} E(y)$

    generalization of y by the variable x

---

[35] The limit $z \leqq [Pr(lx + ly)]^{x+y}$ follows from the fact that each prime number from z is certainly $\leqq$ Pr(lz) and the sum of prime number exponents of z is equal to the sum of the exponents from x and from y, therefore smaller than $x + y$.

16. $0\,Nf\,x = x$

$(n+1)Nf\,x = R(2) \# n\,Nf(x)$

n Nf x formula that arises from *x* through setting the *sign f n* times ahead

17. $Z(n) = n\,Nf[R(1)]$

symbol for number *n*

18. $Typ_1(x) = (Em, n)\{m, n \leqq x \,\&\, [m = 1 \vee 1 Var\, m] \,\&\, x = n\,Nf[R(m)]\}$

sign of type 1

258R

19. $Typ_n(x) = \{n = 1 \,\&\, Typ_1(x)\} \vee \{n > 1$

$\&\, (Ev)[v \leqq x \,\&\, n Var\, v \,\&\, x = R(v)]\}$

sign of type *n*

20. $Elf(x) = (Ea, b, n)\{a, b, n \leqq x \,\&$

$Typ_{n+1}(a) \,\&\, Typ_n(b) \,\&\, x = a \# E(b)\}$

elementary formula

21. $Op(xyz) = x = Neg\, y \vee x = y\, Od\, z \vee$

$(Ev)[v \leqq x \,\&\, Var(v) \,\&\, x = v\, Gen\, y]$

the *formula x* arises from *y* or *z* through the operations of *negation, disjunction, generalization*

22. $F\text{-}R(x) = (n)\{0 < n \leqq l(x) \rightarrow$

$Elf(n\, Gl\, x) \vee (Ep, q)[0 < p, q \leqq n \,\&\, Op(n\, Gl\, x, p\, Gl\, x, q\, Gl\, x)]\}$

259L

$\&\, l(x) > 0$

x is a series of *formulas* each of which is either an *elementary formula* or comes out from the previous ones throughout the operations of *negation, disjunction, generalization*

23. $Form(x) = (En)\{n \leqq Pr[l(x)^2]^{x(lx)^2}$

$\& F\text{-}R(n) \& x = l(n) Gl\, n\}$ [36]

x is a formula (i.e., the last member of a series of formulas)

24. $v\, Fr\, n, x = \ Var(v) \& v = n\, Gl\, x \& Form\, x$

$\& \overline{(Ea, b, c)}[a, b, c \leqq x \& x = a + v\, Gen\, b + c$

$\& Form(b) \& l(a) + 1 < n \leqq l(a) + l[v\, Gen\, b]\}$

the *variable* v is free in x in the n-th place

25. $v\, Geb\, n, x = Var(v) \& Form(x) \& v = n\, Gl\, x \& \overline{v\, Fr\, n, x}$

259R

26. $v\, Fr\, x = (En)[n \leqq l(x) \& v\, Fr\, n, x]$

v occurs in x as a free variable

27. $Su\, x\binom{n}{y} = \varepsilon_z\{(Eu, v) u, v \leqq x \&$

$n = l(u) + 1 \& x = u + R(n\, Gl\, x) + v \&$

$z = u + y + v\}$

$Su\, x\binom{n}{y}$ arises from x when one substitutes y in place of the n-th sign of x

28. $0\, St\, v, x = \varepsilon_n\{n \leqq l(x) \& v\, Fr\, n, x \& \overline{(Ep)}[n < p \leqq l(x) \& v\, Fr\, p, x]\}$

$k + 1\, St\, v, x = \varepsilon_n\{n < k\, St\, v, x \&$

$v\, Fr\, n, x \& \overline{(Ep)}[n < p < k\, St\, v, x \& v\, Fr\, p, x]\}$

$k\, St\, v, x$ is the k-th position (counted from the end of the formula) in which v is free in x

29. $A(v, x) = \varepsilon_n\{n\, St\, v, x = 0 \& n \leqq l(x)\}$

number of positions in which v is free in x

260L

30. $Sb_0(x^v_y) = x$

[36] One recognises the bound $n \leqq Pr[l(x)^2]^{x l(x)^2}$ like this: the length n of a series of formulas that belongs to x can be at most equal to the number of subformulas of x. But there are at most lx subformulas of length 1, at most $lx - 1$ of length 2, and so on, so on the whole at most $lx(lx - 1) < [l(x)]^2$. The prime numbers from n are then all $< Pr(lx)^2$, but their number $< (lx)^2$, and their exponents (which are subformulas of x) $\leqq x$.

$Sb_{k+1}(x^v_y) = Su\{Sb_k(x^v_y)\}\binom{k\,St\,v,x}{y}$

31. $Sb(x^v_y) = Sb_{A(v,x)}(x^v_y)$

 $Sb(x^v_y)$ is identical to the concept $Subst(a^v_b)$ defined above

32. $x\,Imp\,y = N(x)\,Od\,y$

 $x\,K\,y = N[N(x)\,Od\,N(y)]$

 $x\,Aeq\,y = (x\,Imp\,y)\,K\,(y\,Imp\,x)$

 $v\,Ex\,y = N[v\,Gen(N(y))]$

 these correspond to the concepts $\supset, \&, \equiv, (Ex)$

260R

33. $n\,Th\,x = \varepsilon_y(y \leqq x^{(x^n)}\,\&\,(k)\{k \leqq l(x)$
 $\rightarrow [(k\,Gl\,x \leqq 13\,\&\,k\,Gl\,x = k\,Gl\,y) \vee$
 $(k\,Gl\,x > 13\,\&\,k\,Gl\,y = k\,Gl\,x\,[1\,Pr(k\,Gl\,x)]^n)]\})$

 n type elevation of x

34. To the axioms I 1–3 correspond three specific numbers that we designate by $z_1\,z_2\,z_3$ [37]

 $Z\text{-}Ax(y) = [x = z_1 \vee x = z_2 \vee x = z_3]$

35. $A_1\text{-}Ax(y) = (Ea)\{a \leqq x\,\&\,Form(a)\,\&\,x = (a\,Od\,a)\,Impl\,a\}$

 y arises through a substitution in axiom II 1

 A_2Ax, A_3Ax, A_4Ax that correspond to axioms II 2, 3, 4 are defined analogously

36. $A\text{-}Ax(y) = A_1\text{-}Ax(y) \vee \ldots \vee A_4\text{-}Ax(y)$

261L

 y is a propositional axiom

37. $Q(z,y,v) = \overline{(Ep,w)}[v\,Fr\,p,y\,\&\,w\,Fr\,z\,\&\,w\,Geb\,p\,y]$

 z contains no free variable that is bound in a place in y in which v is free

[37] [This item and the next should have x as argument at left.]

38. $L_1Ax(x) = (Ev, y, z, n)\{v, y, z, n \leqq x$
$\& n\, Var(v) \& n\, Typ\, z \& Form(y) \& Q(z, y, v)$
$\& x = (v\, Gen\, y) Impl[Sb(y^v_z)]\}$
x arises from axiom scheme III 1 through substitution

39. $L_2Ax(x) = (E v q r)\{v, q, r \leqq x \&$
$Var(v) \& \overline{v\, Fr\, p} \& Form(p) \& Form(q)$
$\& x = [v\, Gen(p\, Od\, q)] Impl[p\, Od(v\, Gen\, q)\}$
x arises from axiom scheme III 2 through substitution

261R

40. $Red\text{-}Ax(x) = (E u v y n)[u, v, y, n \leqq x \&$
$(n+1) Var(u) \& n\, Var(v) \& \overline{u\, Fr\, y} \&$
$Form(y) \& x = u\, Ex\{v\, Gen[\{R(u) + E(R(v))\} Aeq\, y]\}$
x arises from scheme IV 1 through substitution

50. A determinate number z_4 corresponds to axiom V 1 and we define
$M\text{-}Ax(x) = (En)[n \leqq x \& x = n\, Th\, z_4]$

51. $Ax(x) \sim Z\text{-}Ax(x) \vee A\text{-}Ax(x) \vee$
$L_1\text{-}Ax(x) \vee L_2\text{-}Ax(x) \vee Red\text{-}Ax(x) \vee M\text{-}Ax(x)$
x is an *axiom*

52. $Fl(xyz) = y = z\, Impl\, x \vee$
$(Ev)[v \leqq x \& Var(v) \& x = v\, Gen\, y]$
x is an immediate consequence of y and z

262L

The fact that one can vaguely formulate as follows, namely that each recursive relation (class) can be defined within the system P (meant in a contentful way), without recourse to any contentful meaning of the formulas of P, is expressed in an exact form by the following theorem:

Theorem V.[38] There exists for each recursive relation $R(x_1 \ldots x_n)$ an *n-place relation sign* r (with the free *variables* $u_1 \ldots u_n$) such that

$$(x_1 \ldots x_n) R(x_1 \ldots x_n) \rightarrow Bewb\, Subst(r^{u_1 \ldots\ldots u_n}_{Z(x_1) \ldots Z(x_n)})$$

$$(x_1 \ldots x_n) \overline{R}(x_1 \ldots x_n) \rightarrow Bewb\, Neg\, Subst(r^{u_1 \ldots\ldots u_n}_{Z(x_1) \ldots Z(x_n)})^{39}$$

We rest content to indicate the proof of this theorem in outline, because no difficulties of principle are met in it and because it is rather long-winded. We prove the theorem for all relations $R(x_1 \ldots x_n)$ of the form $x_1 = \varphi(x_2 \ldots x_n)$[40] (in which φ is recursive):

If φ is of the first level, the theorem is trivial. We can therefore use complete induction on the level. Let then φ have level n. It arises in that case from functions of lower levels $\psi_1 \ldots \psi_k$ through the operation of substitution or through recursive definition. By the inductive hypothesis, all is proved for $\psi_1 \ldots \psi_k$, so there exist the associated relation signs $r_1 \ldots r_k$ such that IV [V] holds. The definitional processes

262R

by which φ arises from $\varphi_1 \ldots \varphi_{n-1}$ (substitution and recursive definition) can all be formally reproduced in P. If one does that, one obtains from $r_1 \ldots r_n$ a new relation sign r[41] for which one can prove without difficulty that theorem IV [V] holds, under the condition that it holds for the $r_1 \ldots r_n$.[42] A relation sign r to which a recursive relation is associated in this way (one that contentfully meant expresses that this relation holds) shall

[38] [Changed from IV.]

[39] The variables $u_1 \ldots u_n$ can obviously be given arbitrarily. There exists, for example, always an r with one free variable 11, 13, 17 etc for which ( ) and ( ) hold. [I have added the negation stroke above the lower R, cf. the typewritten version. The empty parentheses have there the numbers 3 and 4 inside.]

[40] From this follows of course at once that it holds for each recursive relation, because such a relation is equal in meaning to $0 = \varphi(x_1 \ldots x_n)$, φ recursive.

[41] In a precise carrying through of this proof, r is not defined through the detour of a contentful meaning, but through a purely formal condition. [Footnote at the bottom of the page, with a circular sign but no corresponding sign in the text. The typewritten version, note 41, gives its place.]

[42] The provability of $Sb(r^{n_1 \ldots\ldots n_n}_{Z x_1 \ldots Z x_n})$ on the basis of the correctness of $R(x_1 \ldots x_n)$ (analogously for negations) depends in the end of course on the decidability of the question whether, for each n-tuple of numbers in a recursive relation, the relation holds and that this decision can be carried through within the system P.

be called *recursive*.

We come now to the goal of our explanations:[43]

Let κ be an arbitrary class of *formulas*. We designate by $Fl(\kappa)$ (set of consequences [Folgerungsmenge]) the smallest set of formulas that contains κ as well as the axioms of group [empty space left] and is closed with respect to the relation of *immediate consequence*. κ is called ω-consistent[44] if there is no recursive class sign a (with the free variable v) such that[45]

$$(n)[Sb(a^v_{Z(n)})\,\varepsilon\, Fl(\kappa)]\,\&\, Neg\, v\, Gen\, a\, \varepsilon\, Fl(\kappa)$$

Then we have the theorem:

263L

The general result about existence of undecidable propositions.

VI There exists for each ω-consistent recursive class κ of [TM: formulas] a recursive class sign a with the variable v such that neither $v\, Gen\, a$ nor $Neg\, v\, Gen\, a$ belong to $Fl(\kappa)$.

Proof: Let κ be an arbitrary recursive ω-consistent class. We define

$$B_\kappa(n) \sim l(n) > 0\,\&\,(x)[x \leqq l(n) \rightarrow$$
$$Ax(x\, Gl\, n) \vee K(x\, Gl\, n) \vee (Ep,q)[p,q \leqq x\,\&\, Fl(x\, Gl\, n, p\, Gl\, n, q\, Gl\, n)]^{46}$$

(cf. the analogous concept $Bew(n)$)

$$x B_\kappa n \sim Bew(x)\,\&\, l(x) = n$$
$$Bewb_\kappa(n) \sim (Ex)\, x\, Bew\, n$$

We clearly have $(n) Bewb_\kappa(n) \sim n\, \varepsilon Fl(\kappa)$

and $(n)\, Bew(n) \rightarrow Bewb_\kappa(n)$

We define now the relation

$$R(xy) \underset{Df}{=} x\, \overline{Bew_\kappa Sb(y\, {}^{13}_{Z(y)})}$$

[43] [There appears for the first time a cursive-looking K used for a class of formulas that I give as κ. The same symbol, possibly drawn smaller, is typeset with no instruction as $\varkappa$ in the printed paper.]

[44] Each ω-consistent κ is of course even so more free from contradiction. The contrary, though, does not hold, as will be shown later.

[45] [Original has as first occurrence of $Fl(v)$, as second $Fl\kappa$.]

[46] [Parentheses after Fl adjusted as in formula (5) of the typewritten version.]

Because $x\,Bew_\kappa\,n$ (by) and Sb (by) are recursive,

263R

then also $R(xy)$. Therefore there exists by theorem V and ( ) a *relation sign* q (with the free *variables* 11, 13) such that the following holds:

$$\overline{x\,B_\kappa Sb(y\,{}^{13}_{Z(y)})} \rightarrow Bew_\kappa Sb(q\,{}^{11}_{Z(x)}\,{}^{13}_{Z(y)})$$
$$x\,B_\kappa Sb(y\,{}^{13}_{Z(y)}) \rightarrow Bew_\kappa Neg\,Sb(q\,{}^{11}_{Z(x)}\,{}^{13}_{Z(y)})$$

We put:

$p = 11Gen\,q$ (p is a *class sign* with the free *variable* 13)

and

$r = Sb(y\,{}^{13}_{Z(y)})$

r is a recursive[47] *class sign* with the free variable 11

Then the following holds:

$$Sb(p\,{}^{13}_{Z(p)}) = Sb(11Gen\,q\,{}^{13}_{Z(p)}) = 11Gen\,Sb(q\,{}^{13}_{Z(p)})$$
$$= 11Gen\,r\,{}^{48} \qquad \text{(because of } \quad)$$

and

$$Sb(q\,{}^{11}_{Z(x)}\,{}^{13}_{Z(p)}) = Sb(r\,{}^{11}_{Z(x)}) \quad \text{(by } \quad)$$

264L

If one now substitutes in formula [missing] p for y, then, taking into consideration ( ) and ( ), there arises:

$$\overline{x\,B_\kappa(11Gen\,r)} \rightarrow Bew_\kappa Sb(r\,{}^{11}_{Z(x)})$$
$$x\,B_\kappa(11Gen\,r) \rightarrow Bew_\kappa N\,Sb(r\,{}^{11}_{Z(x)})$$

Now there results:

1. $11Gen\,r$ is not κ-provable[49] because there is in this case an n so that

[47] For it arises from the recursive *relation sign* q through the replacement of a free variable by a determinate number (p).

[48] The operations *Gen Sb* are naturally always exchangeable (in case they relate to different variables).

[49] x is κ-provable shall mean $Bew_\kappa(x)$ that says the same as $x\,\varepsilon\,Fl(\kappa)$. [Note added at

$n\,Bew_\kappa(11Gen\,r)$. By , $Bew_\kappa N\,Sb(r\,{}^{11}_{Z(n)})$ would then hold, whereas on the other hand, from the κ-provability of $11Gen\,r$ would follow also that of $Sb(r\,{}^{11}_{Z(n)})$. So κ would be inconsistent (and the more so ω-inconsistent).

2. $Neg 11Gen r$ is not κ-provable. For were this the case, then (in accordance with the ω-consistency of κ), $Sb(r\,{}^{11}_{Z(n)})$ would not be κ-provable for all n. It follows from this by that there is in this case an n such that $n\,Bew_\kappa(11Gen\,r)$, i.e., $11Gen\,r$ would be provable, which is impossible because of the freedom from contradiction of κ. Then $11Gen\,r$ is undecidable from κ by which theorem VI has been proved.

[At this point, an addition is indicated. The addition is found in pages ripped off the manuscript, microfilmed in reel 20, frame 496. The same text is found on page 31 of the typewritten manuscript.]

If one adjoins $Neg 11Gen\,r$ to κ one obtains a consistent but not ω-consistent class κ'. κ' is consistent, because otherwise $11Gen\,r$ would be κ-provable. κ' is, though, not ω-consistent because $\overline{Bewb_\kappa 11Gen\,r}$ holds, in other words $\overline{(x)xB_\kappa(11Gen\,r)}$, and then there follows [added below the text: on the basis of lemma] $(x)Bew_\kappa Sb(r\,{}^{11}_{Z(x)})$, and even the more so $(x)Bew_{\kappa'} Sb(r\,{}^{11}_{Z(x)})$, and on the other hand there holds naturally $Bew_{\kappa'} Neg\,Gen\,r$. [End of addition.]

264R

A special case of theorem VI is the one in which the class κ consists of finitely many axioms α (and those infinitely many ones that arise from them by type elevation). Each finite class α is naturally recursively definable. Let M be the greatest number contained in α. Then in this case the following holds for κ:

$$\kappa(x) \sim (En,m)[m \leqq x \,\&\, n \leqq M \,\&\, \alpha(n) \,\&\, x = m\,Th\,n]$$

Then also κ is recursively definable. This allows one to conclude, for example, that not all problems are solvable even with the help of the axiom of choice (for all types) or with the generalized continuum hypothesis (assuming that these hypotheses are ω-consistent).[50]

bottom of page, placed here as in the TM.]

[50] From the fact that one never obtains a system definite with respect to decision through the addition of finitely many axioms, it follows especially that each formal system that falls under theorem V has the ordinal degree of incompleteness Ω (in the sense of). [This note, with the name A. Tarski, is cancelled in the typewritten version.]

In the proof of theorem V [VI], no other properties of the system P need be used than the following:

1. The class of axioms and the rules of inference (i.e., the relation of immediate consequence) are recursively definable.

2. Each recursive relation can be defined inside the system (in the sense of theorem IV).

Therefore *there exist in each formal system* in which conditions 1 and 2 are satisfied and that is ω-consistent undecidable propositions of the form $(Ex)F(x)$ in which F is recursive (and the same for each extension of such a system by a

265L

recursively definable class of axioms).

To systems that satisfy conditions 1–2 belong, as is easy to ascertain, especially the axiom systems of set theory of Zermelo Fraenkel and of von Neumann.[51]

We derive now further consequences from theorem VI and give to this purpose the following definition:

A relation (class) is called *arithmetic* if it can be defined with the help of just the concepts $+ \times$ (addition and multiplication applied to natural numbers) and the logical constants $\vee \; \bar{} \; (\;) \; =$, in which $(\;)$ and $=$ are allowed to apply only to natural numbers.[52] The concept of an arithmetic proposition is defined correspondingly. In particular, the relations "greater than" and "congruent" for example are arithmetic, because we have:

$$x \equiv y \overset{mod}{(n)} \equiv (Ez)(x = y + zn \vee y = x + zn)$$

$$x > y \equiv (Ez)\, y = x + z \quad \text{[TM has existence negated]}$$

The following theorem holds:

[51] The proof of condition 1 presents itself even as simpler than in the case of system S, because there is only one kind of ground variable and one basic relation ε. The axioms run pretty much parallel to the above.

[52] The *Definiens* of such a concept must be built with only the help of the signs introduced and variables for entire numbers $x\,y$ (there must not occur function variables).

265R[53]

VI [VII] Each recursive relation is arithmetic.

We prove the theorem in the form: Each relation of the form $x_0 = \varphi(x_1 \ldots x_n)$ in which φ is recursive, is arithmetically definable, and we use complete induction on the level of φ. Let φ have level n. We have either:

1. $\varphi(x_1 \ldots x_n) = \varrho(\chi_1(\mathfrak{x}_1) \ldots \chi_n(\mathfrak{x}_n))$ [54]

Here ϱ and all of the χ_i have levels lower than n.

Or we have:

2. $\varphi(0 x_2 \ldots x_n) = \psi(x_2 \ldots x_{n-1})$

$\varphi(k{+}1\, x_2 \ldots x_n) = \mu(k\, \varphi(k\, x_2 \ldots x_n))$

in which ψ μ are of a lower level than n.

In the first case, we have[55]

$x_0 = \varphi(x_1 \ldots x_n) \sim (Ey_1 \ldots y_n)[R(x_0\, x_1 \ldots x_n)\ \&$

$S_1(x_1 \mathfrak{x}_1) \ldots \& \ldots S_n(x_n \mathfrak{x}_n)]$

266L

in which R and S_i are the arithmetic relations that, by the inductive assumption, exist and are equivalent to

$y = \varrho(x_1 \ldots x_n)$ and $y = \chi_i(\mathfrak{x}_i)$, respectively

Therefore $x_0 = \varphi(x_1 \ldots x_n)$ is in this case arithmetic.

In the second case, we use the following procedure: One can define the relation $y = \varphi(x_1 \ldots x_n)$ with the help of the concept of a "sequence of numbers" (f), in the following way:

$$(Ef)\{f_0 = \varphi(0\, x_2 \ldots x_{n-1})\ \&\ (k)[k < x_1 \rightarrow f_{k+1} = \mu(k\, f_k)]\ \&\ f_{x_1} = y\}$$

or when S and T are the relations that, by the inductive assumption, exist and are of a kind equivalent to, respectively, [TM: $y = \psi(x_2 \ldots x_n)$] and

[53] We use German letters $\mathfrak{x}_i$ as abbreviations of n-tuples of variables $x_1 \ldots x_n$.

[54] [The arguments are close to illegible, perhaps changed from the German $\mathfrak{x}_1 \ldots \mathfrak{x}_n$ to $\mathbf{x}_1 \ldots \mathbf{x}_i$. The typewritten version has $\chi_1(x_1 \ldots x_n)$ etc.]

[55] [Gödel changed the bound variables from x_i to y_i in the quantifier but left those in the relation unchanged.]

[TM: $z = \mu(x_2 \ldots x_{n+1})$], through[56]

$$(Ef)\{Sf_0x_2 \ldots x_n \,\&\, (k)[k < x_1 \rightarrow f_{k+1} = \mu(k\, f_k)] \,\&\, f_{x_1} = y\}$$

[Bottom of the page has the undesignated footnote: We denote by f_x the x-th member of the sequence x [should be f].]

266R

We replace now the concept "sequence of numbers" by the concept "pair of numbers," by associating to the pair n, d the sequence of numbers

$$f^{nd} \;\{f_k^{nd} = [n]_{1+(k+1)d}\}$$

in which $[n]_p$ denotes the smallest non-negative remainder of n modulo p.

We have then the

Lemma 1. If f is an arbitrary sequence and k an arbitrary natural number, there exists a pair of natural numbers $n\, d$ such that f^{nd} and f agree on the k first members.

Proof. Let l be the greatest of the numbers $k\, f_0\, f_1 \ldots f_{k-1}$. n is determined so that

$$n \equiv f_i(1 + (i+1)l!) \;\text{ for } i = 0, 1, \ldots k - 1$$

something that is possible because each two of the numbers

$$1 + (i+1)l! \;\; i = 0, 1 \ldots k - 1$$

are relatively prime.[57] The pair of numbers $n, l!$ fulfils therefore what is required.

267L

The relation $x = [n]_p$ is defined by

$$x \equiv n(p) \,\&\, x < n$$

It is therefore arithmetic, and then also the relation

$$(Emd)\{S([m]_{d+1}0\, x_2 \ldots x_n) \,\&$$

$$(k)[k < x_1 \rightarrow T([m]_{d(k+2)+1}k[m]_{d(k+1)+1}]$$

---

[56] [TM has:
$x_0 = \varphi(x_1 \ldots x_n) \sim (Ef)\{S(f_0, x_2 \ldots x_n) \& (k)[k < x_1 \rightarrow T(f_{k+1}, k, f_k, x_2 \cdots x_n)] \& x_0 = f_{x_1}\}$]

[57] A prime number contained in two of these would also have to be contained in the difference $(i_1 - i_2)l!$ and therefore, because of $i_1 - i_2 < l$, in $l!$.

$\& y = [m]_{d(k_1+1)+1}\}$

that is by Lemma 1 equivalent to $y = \varphi(x_1 \dots x_n)$. (With the sequence f, the question is only about its course of values up to the k-th member.) Hereby the lemma is proved.

In accordance with theorem VI [VII], there exists for each problem of the form $(En)F(n)$ (F recursive) an equivalent arithmetic problem, and because the whole proof of theorem VI [VII] can be formalized within the system P, this equivalence is also provable. Therefore we have:

Theorem VI [VIII] There exist in each of the formal systems[58] mentioned in theorem V [VI] undecidable arithmetic propositions (i.e., ones in which only the concepts $+$ $\times$ applied to natural numbers occur).

20-496 [59]

The same holds for axiom systems of set theory and their extension by ω-consistent recursive classes of axioms.

[There follows the addition to page 264L.]

For the existence of undecidable propositions, it is sufficient just to assume for κ next to ω-consistency the following: There exists a class sign r such that $\kappa(x) \rightarrow BewSb(r^{\,n}_{Z(x)})$ and $\overline{\kappa}(x) \rightarrow BewSb(r^{\,n}_{Z(x)})$. In other words, it is decidable for each number in P whether it belongs to κ. For this property carries over from κ to B_κ, and only this is required in the above proof.

284L[60]

We derive finally the following result:

Theorem. There exist in all of the formal systems mentioned in theorem VI unsolvable problems within the narrower functional calculus.

---

58 [Footnote found on page 20-496: I.e., in those ω-consistent systems that arise from P through the addition of a recursively defined class of axioms.]

59 [This page belongs to a loose sheet found in reel 20. It has been clearly ripped off from the notebook, because "black holes," revealed as ink spots, match to perfection such spots on pages 267L and 267R. The text continues the previous in the same way as the typewritten version. There is on these loose pages also an addition to several pages before, on page 264L, and more remarks that are difficult to locate precisely.]

60 [The following four pages are written in backward direction from the end of the notebook. Their contents can be profitably compared with theorem X and its proof with which the original version of Gödel's article ends.]

283L

This depends on the following:
Theorem N [TM: X]. Each problem of the form $(x)\varphi(x)$ in which φ is a recursive property can be reduced back to the satisfiability of a formula of the narrower functional calculus.[61]

We count as formulas of the narrower functional calculus (eF) those formulas that are built up from the basic signs $(\ )\ ^{-}\ \vee = x\, y$ (individual variables) and class and relation variables $F(x)\ G(x\,y)$ in which $(\ )$ and $=$ [62] are allowed to apply only to individuals. We join to these signs even a third kind of variables $\varphi(x)\ \psi(xy)\ \chi(xyz)$ that represent functions over objects (i.e., $\varphi(x)\ \psi(xy)$ denote (unique) functions the arguments and values of which are individuals).

283R

We shall call a formula that contains, beyond the basic signs of the eF introduced above, variables of the third kind $\varphi\ \psi$ etc a formula in the extended sense (iwS) [im weiteren Sinne].[63] The concepts satisfiable, generally valid carry over without further into formulas in the extended sense. The theorem holds that gives, for a formula iwS ($\mathfrak{A}$), a usual formula $\mathfrak{B}$ of the eF such that the satisfiability of $\mathfrak{A}$ is equivalent to the satisfiability of $\mathfrak{B}$. $\mathfrak{B}$ is obtained from $\mathfrak{A}$ through the replacement of function variables $\varphi(x_1\, y\, x_2)$ that occur in $\mathfrak{A}$ through expressions of the form $(\imath y)F(y\, x_1 \ldots x_2)$, i.e., $\imath$ the descriptive function $(\imath x)(\ ,\ )$ in the sense of the *PM*, is resolved [TM has I∗14, a reference to the *PM*] and

282L

the formula obtained in this way logically multiplied by an expression that states that all of the F put in place of φ are unique in respect of the first

[61] Cf. Hilbert Ackermann *Grundzüge*.

[62] One can, however, give to each formula in which the sign = occurs another one in which it doesn't occur anymore, the satisfiability of which is equivalent to the satisfiability of the original formula. Therefore theorem N holds also for the narrower functional calculus in the sense of Hilbert and Ackermann. Formulas with the sign = are not counted among the narrower functional calculus in Hilbert Ackermann.

[63] Variables of the third kind are allowed to stand everywhere as placeholders for individual variables, for example $F(\varphi(x)), \varphi(x) = y$, and substituted arbitrarily inside each other, for example $\varphi[\psi(\chi(x), y)z]$.

placeholder.

We show now that there is to each problem of the form $(x)\varphi(x)$ [κ written above φ] (φ recursive) an equivalent one as concerns the satisfiability of a formula iws, from which theorem N follows by the remark made. Since κ is recursive, there is a recursive function φ such that $\varphi(x) = 0 \sim \kappa(x)$, and there is for φ a series of functions $\varphi_1 \dots \varphi_n$ such that $\varphi_n = \varphi$, $\varphi_1 = f(x)$ (successor function) and for each k either:

1.) $(x_1 \dots x_n)\psi_k(0\, x_2 \dots x_n) = \psi_l(x_2 \dots x_n)$ (1)

$(x\, x_1 \dots x_n)\psi(\psi_1(x), x_2 \dots x_n) = \varphi_n(x, \psi(x \dots x_n))$

$l, n < k$

2.) or $(x_1 \dots x_n)\varphi_k(x_1 \dots x_n) = \varphi_l(\varphi_{i_1}(x_1) \dots \varphi_{i_n}(x_n))$ (2)

The theorem holds in the case that φ_n is a constant function that forms the negation.

20-495L[64]

$P(x_1) \sim (Ex, y)\{$

$P(x_2, x, y) \sim (Ex_1 y_1)\{(v_1)x_2(v_1) \sim v_1 = x_1 \vee v_1 = y_1\ \&$

$(x, y)[$

65

$G(x, y, x_1) = (v)[x_1(v) \sim x = v \vee y = v]$

$G(x, x_1) = (v)[x_1(v) \sim x = v]$

$P(x_2, x_1, y) \sim (Ex, y_1)[G(x_1 y_1 x_2)\ \&\ G(x_1 x)\ \&\ G(y_1 xy)]$

$R(x_3, xy) = (Ex_2)[x_3(x_2)\ \&\ G(x_2 x\, y)]$

$Eind(x_3) \sim (x)(Ey)(v)[R(x_3 x\, v) \sim v = y]$

$P(z_0 x_1 \dots x_{n-1})$

$Q(z_0 x_1 \dots x_{n+1})$

$(Ex_3)Eind(x_3)\ \&\ [(x)x_3(0\, x) \rightarrow P(x_1 x_2 x_3 \dots x_n)$

[64] [This seems an inconclusive page with notation and definitions that continue on page 20-495R, in preparation of the theorem on page 284L that begins the backward direction of the *Heft*. Comma and subscript 1 are indistinguishable as written. Compare also page 335R for *Eind*.]

[65] [Squeezed between the lines here: $G(x_1 y_1 x_2)$]

$(y)(x)(k)[x_3(k\,x)\,\&\,x_3(k+1\,y) \rightarrow Q(y,k,x,x_2\ldots x_n)\,\&\,x_3(x_1,x_0)]$

20-495R[66]

$$Q:\ (x)\overline{\varphi_1(x) = x_0}\,\&\,(xy)\{[\varphi_1(x) = \varphi_1(y)] \rightarrow x = y\}$$
$$R:\ (x)\varphi_n(x) = 0$$

The formula $(Ex_0)\{P\,\&\,Q\,\&\,R\}$ (abbreviated designation Z) has then the required property, i.e.

1. If $(x)\Phi(x) = 0$ holds, Z is satisfiable, because the functions $\Phi_i(1 = 1\ldots k)$ clearly give when substituted in Z a correct proposition.
2. If Z is satisfiable, $(x)\Phi(x)$ holds.

Proof. Let Ψ_i be the functions that exist by assumption and that deliver, when substituted in Z, a correct proposition. Their domain of objects is $\mathfrak{J}$. We designate by $(Ex_0)P'\&Q'\&R'$ the proposition that arises through substitution of the Ψ_i from $(Ex_0)P\&Q\&R$, and by a_0 $\{a_0\,\varepsilon\,\mathfrak{J}\}$ one of the individuals x_0 the existence of which is claimed in this proposition. We build now

267R

the smallest subclass of $\mathfrak{J}$ that contains a and is closed with respect to the operation $\Phi_1(x)$. This subclass ($\mathfrak{J}'$) has the property that each of the functions Ψ_i, when applied to an element from $\mathfrak{J}'$, gives again an element of $\mathfrak{J}'$, for this follows for Ψ_1 from the definition, and because of the correctness of the formulas [(1), (2), (3)] for Ψ_i, this property is carried over from formulas with a lower index to ones with a higher one. We call Ψ_i' the functions that arise from the Ψ_i through a limitation to the domain of individuals $\mathfrak{J}'$. The individuals from $\mathfrak{J}'$ can be (because of the correctness of Q' for $x_0 = a$) mapped in a unique way to the non-negative integers and moreover so that a goes over to 0 and the function Ψ_1' to the successor function. By this mapping, the function Ψ_i' goes, then, over to the function Φ_i and because of the correctness of R', $(x)(\Phi_n(x) = 0)$, or $(x)\,\Phi(x) = 0$ holds, as was to be proved.

The considerations that have led to the proof of theorem [TM: IX] can

[66] [This "black hole" page continues with page 267R. The backside of this page is page 20-496 that has additions to the text in different places, as indicated above.]

be carried through also within the system P. Therefore the equivalence between a proposition of the form $(x)F(x)$ (F recursive) and the satisfiability of the corresponding formula of the narrower functional calculus is provable in P. From the undecidability of the one proposition follows that of the other, by which theorem X has been proved.[67]

268L

To finish, let us point at the following interesting circumstance that concerns the undecidable proposition S put up in the above. By a remark made right in the beginning, S claims its own unprovability. Because S is undecidable, it is naturally also unprovable. Then, what S claims is correct. We have, then, decided with the help of metamathematical considerations a proposition S that is undecidable in the system. The precise analysis of this state of affairs leads to interesting results that concern a proof of freedom from contradiction of the system P (and related systems) that will be treated in a continuation of this work soon to appear.

[The rest of the page indicates the following additions to the manuscript, lightly cancelled which should just mean that they have been incorporated in the text:]

Footnote p. 1. Cf. the summary of results of this work that appeared in [the place of publication of the 1930 note is added here in the proofs of the article].

Footnote By formulas of the narrower functional calculus of the system P are naturally to be understood those that arise from formulas of the *PrM* through the replacement of relations by classes of higher types, as indicated on page footnote p. 12 [This passage is found in the typewritten manuscript, and in the final article as footnote 54, p. 193.]

268R [The two titles below are followed by recursion equations for the predecessor operation, recursion schemes, and schemes for the ε-operator.]

On the existence of undecidable mathematical propositions in the sys-

[67] It follows from theorem X that, for example, Fermat's theorem and that of Goldbach would become decided if one had solved the decision problem of the narrower functional calculus, because both of these problems can be easily brought into the form $(x)F(x)$ (F recursive).

tem of the *Princ math*

On unsolvable mathematical problems in the system of the *Principia Mathematica*

269 [The left page of this frame is blank, the right side has just formulas with the ε-operator.]

270 [The left page has recursion schemes, the right a plan of contents for a lecture on completeness. Gödel gave such a lecture in Vienna on 28 November 1930.]

1. How do the building blocks of an axiom system look like
   basic objects, basic relations
2. How do the axioms look like
   a.) Logical signs, meaning explanation
   b.) Axioms of the first kind
   c.) Axioms of the second kind
   d.) Reduction of one to the other
3. How is the construction carried out
   counterexample, inference, inference formalizable by Frege, problem of completeness
4. Transition to the narrower functional calculus
   Setting out of the problem by Hilbert
5. Setting up of the axioms and rules of inference
6. The problem for the propositional calculus and a brief solution
7. Production of normal form
8. Formula of grade *k*
9. Formula of first grade

271 [The left page has four equations for higher-level functions, followed by unrelated computations in red pencil that continue to the next page.]

8. Let us turn back to the undecidable proposition

272L

Dear Mr von Neumann

Many thanks for your letter of [20 November]. Unfortunately I have to inform you that I have been in possession of the result you communicated since about three months. It is also found in the attached offprint of a communication to the Academy of Sciences. I had finished the manuscript for this communication already before my departure for Königsberg and had presented it to Carnap. I gave it over to Hahn for publication in the *Anzeiger* of the Academy on 17 September. [Cancelled: The reason why I didn't inform you in any way [written heavily over: didn't tell anything] of my second result in Königsberg is that the precise proof is not suited to oral communication and that an approximate indication could easily arouse doubts about [heavily cancelled: correctness] executability (as with the first) that would not appear convincing.] As concerns the publication of this matter, there will be given only a shorter outline of the proof of impossibility of a proof of freedom from contradiction in the *Monatsheft* that will appear in January [changed into: early 1931] (the main part of this treatise will be filled with the proof of existence of undecidable sentences). The detailed carrying through of the proof will appear in a *Monatsheft* only in July or August. I can send you a copy [Abschrift] of – proofs of my next work in a few weeks. [Cancelled: I would be readily prepared, though, to send you already the manuscript – a copy of the manuscript. I am, though, [prepared] to send you a part of my work that

272R

relates to freedom from contradiction, in a few days as a manuscript, so that you see to what extent there obtains accordance with my result.]

I shall include the part of my work that concerns the proof of freedom from contradiction in a manuscript, so that you can see from it to what extent your proof matches mine.

The carrying through of the proof appears together with my proof of undecidability in the next volume of the *Monatshefte*. I didn't want to talk about it before a publication because this thing (even more than the proof of undecidability) must arouse doubt about its executability before it is laid out in an exact way.

273L

Let us now turn back to the undecidable proposition 17*Gen r*. We shall denote the proposition that "κ is free from contradiction" by $Wid(\kappa)$. For the proof of the theorem that 17*Gen r* is unprovable, only the freedom from contradiction of κ was used (cf. 1.) on page 30). So we have

$$Wid(\kappa) \rightarrow \overline{Bew_\kappa}(17Gen\, r)$$

therefore by (6·1) $\quad Wid(\kappa) \rightarrow (x)[\overline{x\, B_\kappa}(17Gen\, r)]$

and by (15) $\quad Wid(\kappa) \rightarrow (x)Bew_\kappa[Sb\, r\, {}^{17}_{Z(p)}]$

By $\quad$ we have $\quad r = Sb(q\, {}^{19}_{Z(p)})$

and q is by $\quad$ the relation sign (with the variables 17, 19) that, meant contentfully, expresses that $R(xy)$,

$R(xy) \sim \overline{x\, B_\kappa Sb(y\, {}^{17}_{Z(y)})}$ holds

Therefore r expresses that

$\overline{x\, B_\kappa Sb(p\, {}^{17}_{Z(p)})} \quad \longrightarrow \quad \overline{x\, B_\kappa\, 17Gen\, r}$ holds.

We designate this by $R'(x)$

273R

and by $\quad$, we have

$R'(x) \rightarrow Bew_\kappa Sb(r\, {}^{17}_{Z(x)})$

$\overline{R'(x)} \rightarrow Bew_\kappa Neg(r\, {}^{17}_{Z(x)}$ $\quad$ [incomplete]

(for even r is recursive), that is in other words, r is, contentfully meant, the class $R'(x)$. Therefore we have:

$Wid(\kappa) \rightarrow (x)R(x)$ [1]

It is easy to show that all concepts that occur in section 2 (beginning with the concept recursive included) are definable within the system P and all

---

[1] One has to think for all these considerations a recursive class κ as a basis, chosen once and for all (but arbitrarily) (thereby also q, r, p become fixed for values that can be given) (the simplest assumption is that κ is the empty set).

theorems proved about them even provable in P.[2]

Let, especially, w be the *formula* that exists by the remark made and that denotes, contentfully meant, $Wid(\kappa)$. Since r denotes, contentfully meant, $R'(x)$,

274L

then $(x)R(x)$ becomes expressed by 17*Gen r*. We have therefore [added between lines: and because $Wid(\kappa) \rightarrow (x)R(x)$ is, by a remark already made, provable in P, i.e.], the proposition [added in margin: exactly expressed the propositional sign] w *Impl* 17*Gen r* is *provable*, and because (even the more κ provable[3]) 17*Gen r* is by the above [not] κ provable, then also w cannot be κ provable,[4] i.e., the freedom from contradiction of κ (with arbitrary recursive κ free from contradiction) is unprovable.

Even this proof is constructive, i.e., if a formal proof of freedom from contradiction were at hand, one could effectively construct a contradiction in κ. The theorem can be very easily generalized (the theorem about undecidability), like this.

274R

It can be, for example, extended word for word to set theory and to axiom systems for classical mathematics.[5] In each of these systems S, the (proposition expressible therein) A that states that S is free from contradiction is unprovable in S. The proof given in the last section need not be carried out in all details. A completion in this direction as well as precise characterisation of systems in which on the one side [sentence breaks off but the preceding cancelled sentence reads: A completion in this direction as well as precise characterisation of systems in which it holds is to follow next to other results in a continuation of this work.]

We have shown above that through the adjunction of *Neg*17*Gen r* to κ, a

[2] One can convince oneself about this step by step, for the concepts defined and theorems proved from page on.

[3] [Parenthetical remark probably added afterwards at the end of line. An improved formulation is found on page 276L.]

[4] [Changed from: because 17*Gen r* is by the above undecidable, then also w must be unprovable.]

[5] Cf. J. v. Neumann [TM: Zur Hilbertschen Beweistheorie].

system arises that is not ω-consistent. Therefore one can (roughly speaking) infer that *Neg*17*Gen r* is false, consequently 17*Gen r* correct. We have, after all, decided a proposition undecidable from κ. When one analyses precisely what has been proved, it turns out that [sentence breaks off]

275L [This page has an addition to page 276L.]

275R[6]

From the outcome of section 2, a strange result follows that concerns proofs of absence of contradiction of the system *P* (and its extensions). It is stated by the following theorem. Let κ be an arbitrary recursively defined class of formulas free from contradiction. Then we have:

Theorem IX The proposition (expressible in *P*) that P_κ[7] is free from contradiction is not κ-provable. Or: The freedom form contradiction of *P* (and its extension P_κ) becomes unprovable in *P* (P_κ), on the condition that *P* (P_κ) are free from contradiction. In the contrary case, each proposition is obviously provable.

Proof: Let κ be an arbitrary (one that remains fixed in the considerations to follow) recursive class (in the simplest case an empty class). To prove the fact that 17*Gen r* is not κ-provable one needs, as emerges from the above page , to use only the freedom from contradiction of κ, i.e., we have:

$$Wid(\kappa) \rightarrow \overline{Bew_\kappa}(17Gen\,r)$$ [8]

or by (6·1) $$Wid(\kappa) \rightarrow (x)x\overline{B}_\kappa(17Gen\,r)$$

276L

and therefore:

$$Wid(\kappa) \rightarrow (x)x\overline{B_\kappa}Sb(p^{19}_{Z(p)})$$

or by $$Wid(\kappa) \rightarrow (x)Q(x,p)$$

[6] [The following three pages are almost identical to section 4 of Gödel's published paper, with just an added remark about Hilbert's program. The missing cross-references can be gathered from the latter.]

[7] We designate by P_κ the system that becomes provable from *P* after the adjunction of axiom class κ.

[8] $Wid(\kappa) = (Ex)[Form(x) \,\&\, \overline{Bew_\kappa}(x)]$.

We ascertain now the following:[9] All of the concepts defined so far in sections 2[10] and 4 (and all of the claims proved) are also definable (provable) in P. For we have applied the usual methods of definition and proof as these are formalized in system P.

Let w be the sentence formula through which $Wid(\kappa)$ is expressed in P.

$Q(xy)$ is correspondingly expressed through q (in Z-ordinals),

$Q(xp)$ consequently through r $(= Sb(q\,{}^{19}_{Z(p)})) = r$,

and $(x)Q(xp)$ through $17Gen\,r$.

Because of , $w\,Impl(17Gen\,r)$ is then provable in P (even the more provable from κ). Were now w κ-provable, then also $17Gen\,r$ would be κ-provable, and from this would follow by that κ is not free from contradiction, by which theorem IX is proved.

Let it be remarked that even this proof is constructive, i.e., it admits, in case w were proved, a contradiction to be shown in κ.

The whole proof lets itself be carried over word for word to set theory (M) or to analysis (A). For that, one needs just to replace the metamathematical concepts that concern P, defined on page ,

276R

by the corresponding concepts about M and A. The result would be the same as in theorem IX: There exists no proof of freedom from contradiction for A (M) that could be formalized within A (M), on the condition that A (M) are free from contradiction.[11]

We have limited us in this work, in section 2 as well as 4, essentially to the system P, and have indicated only in outline the application to other systems. The results will be expressed and proved in full generality in a continuation of this work soon to follow. In this work, even the proof of XI, conducted here in a somewhat sketchy fashion, will be presented in detail.

277L

Dear Mr von Neumann!

---

[9] [Addition from page 275L.]

[10] From the definition of recursive up to the proof of theorem VIII [TM note 66: VI] inclusive, [?] cf. J. v. Neumann.

[11] [The typewritten manuscript has here the remark from page 279R.]

Hearty thanks for your letter of 20·/IX [XI]. The result of which you write to me is known to me since already about three months, but I didn't want to talk anything about it before I had brought the proof into a print-ready form. I send you enclosed an offprint in which the mentioned theorem gets expressed. I had finished the manuscript of this communication to the Academy already before my departure to Königsberg and presented it to Carnap. I gave it to Hahn for publication in September. The carrying through of the proof will appear together with the proof of undecidability in a near *Monatsheft* (beginning of 1931). I shall have proofs of this work in a few weeks and will then send them to you immediately.

[Lightly cancelled: I have limited myself in this work, following the main issue, to the system of *PM*, and as concerns other systems, remained content with the indication that the proof can be carried through similarly. I wanted to prove in detail the general result to this effect only in a

277R

continuation of this work.]

Now to the matter itself. The basic idea of my proof can be described (quite roughly) like this. The sentence *A* that I have put up and that is undecidable in the formal system *S* asserts its own unprovability and is therefore correct. If one analyses precisely how this undecidable sentence *A* could still be metamathematically decided, it appears that this became possible only under the condition of the freedom of contradiction of *S*. That is, it was strictly taken not *A* but $W \rightarrow A$ that was proved (*W* means the proposition: *S* is free of contradiction). The proof of $W \rightarrow A$ lets itself be carried through, though, within the system *S*, so that if even *W* were provable in *S*, then also *A* which contradicts the undecidability of *A*.

[Cancelled: As concerns the meaning of this result, my opinion is that it does not reach as far as you suggest in your letter, namely that, as you write, "the unprovability of freedom from contradiction of mathematics " would have been proved.]

278L

As concerns the meaning of this result, then, my opinion is that *only* the impossibility of a proof of freedom from contradiction for a system *within the system itself* is thereby proved. (I.e., one cannot pull oneself up by one's own bootstraps from the swamp of contradiction.) For the rest, I am fully convinced that there is [cancelled: a finite] an intuitionistically unobjectionable proof of freedom of contradiction for classical mathematics [added above: and set theory], and that therefore the Hilbertian point of view has in no way been refuted. Only one thing is clear, namely that this proof of freedom from contradiction has in any case to be far more complicated than one had assumed so far.

As concerns the question of translation, then, I don't share your opinion, but my opinion is instead that there exists no formal system in which all [cancelled: intuitionistically unobjectionable constructive] finite proofs would be expressible*[12] (and even [cancelled: constructive] finite in the strictest sense, i.e., without choice sequences)

278R

would be expressible. Still, I would like very much to hear about your contrary argument concerning the matter. I would be further interested in whether your proof is built on the same thoughts as mine, concerning which, from what you intend in relation to publication, namely that you refer to my work in yours, would be something I very much hope in any case.

Unfortunately, nothing seems to come of my travel to Berlin this year.

In the hope of a swift reply, I remain with

best wishes, yours sincerely

279L [blank]

279R [This remark is a part of Gödel's section 4 in the printed paper.]

Let it be expressly stated that theorem IX (and the corresponding results about A, M) stand in no contradiction with the Hilbertian formalistic standpoint. For the latter requires only the existence of a finite consistency proof and it is not at all excluded that each finite proof has to be representable in $P\ (A, M)$.

---

12 [The asterisk directs to an addition at the end of the letter draft:] * From the treatise of P. Bernays on "Philosophie der Mathematik und die hilbertsche Beweistheorie" in the *Blätter für Deutsche Philosophie*, volume 4, issue 3/4, 1930, I see that this is also the view of Hilbert and Bernays (cf. what is said on page 366).

Part IV

The typewritten manuscripts

J. von Plato, *Can Mathematics Be Proved Consistent?*, Sources and Studies in the History of Mathematics and Physical Sciences, https://doi.org/10.1007/978-3-030-50876-0_4

Some metamathematical results on definiteness with respect to decision and on freedom from contradiction

by

Kurt Gödel, Wien.

-1-

If one builds on top of the Peano axioms the logic of the Principia mathematica[1] (natural numbers as individuals), with the axiom of choice (for all types), a formal system S arises for which the following theorems hold:

I. The system S is *not* definite with respect to decision, i.e., there exist within the system propositions A (and such can even be given) for which neither A nor $\overline{A}$ is provable, and there exist even some undecidable problems of the simple structure: $(Ex)F(x)$ in which x runs over the natural numbers and F is a property of natural numbers (even definite with respect to decision).

II. Even if one allows all the logical means of the Principia Mathematica in metamathematics (especially the extended functional calculus[1] and the axiom of choice), there is *no proof of freedom from contradiction* for the system S (even the less if one restricts the means of proof in some way). A proof of freedom from contradiction of the system S can, then, be carried through only by auxiliary means that lie *outside* the system S, and the case is analogous also for other formal systems, say the Zermelo-Fränkel axiom system of set theory.

III. Theorem I can be sharpened in that even if finitely many axioms are added to the system S (or infinitely many that arise from the finitely many through "type elevation"), *no* system definite with respect to decision arises as soon as the

-2-

extended system is $\aleph_0$-consistent. Here, a system is called $\aleph_0$-consistent if for no property of natural numbers $F(x)$, we have as simultaneously provable:

[1] With the axiom of reducibility or without ramified type theory.

$$F(1), F(2) \ldots F(n) \ldots \quad \text{ad inf.}$$

and $(Ex)\overline{F(x)}$. (There exist extensions of the system S that are consistent but not $\aleph_0$-consistent.)

IV. Theorem I holds even for all $\aleph_0$-consistent extensions of the system S by *infinitely many* axioms, as soon as the class added is definite with respect to decision, i.e., it is decidable for each formula whether it is an axiom or not (herein, the logical means of the Principia Mathematica are again assumed.)

Theorems I, III, IV, can be extended also to other formal systems, for example the Zermelo-Fränkel axiom system of set theory. [Added by Gödel.]

The proofs of these theorems will appear in the *Monatshefte für Mathematik und Physik*. [Added by Hahn.]

- 1 -

On formally undecidable propositions of Principia Mathematica and related systems[1]).

By KURT GÖDEL in Vienna.

1.

The development of mathematics in the direction of greater exactness has led, as is well known, to wide areas of it having been formalized, in a manner in which proofs can be carried through by a few mechanical rules. The most comprehensive formal systems put up at the time are on the one hand the system of the Principia Mathematica (PM)[2]), on the other hand the axiom system of set theory[3]) of Zermelo–Fraenkel (developed further by J. v. Neumann). Both of these systems are so comprehensive that all the proof methods used in mathematics today are formalized in them, i.e., led back to a few axioms and rules of inference. Therefore,

- 2 -

the conjecture lies close at hand that these axioms and rules of inference are sufficient to carry through any proof at all thinkable. It will be shown in what follows that this is not the case but that there exist in both of the systems put forward even relatively simple problems from the theory of ordinary entire numbers[4]) that cannot be decided from the axioms. It is a

---

1) Cf. the summary of results of this work that appeared in [the Anzeiger der Akad. Wiss. in Wien (math.-naturw. Kl.) 1930, Nr. 19.]

2) We count among the axioms of system PM especially also: the axiom of infinity (in the form: there exist exactly denumerable many individuals), the axiom of reducibility, and the axiom of choice (for all types).

3) Cf. A. Fraenkel, Zehn Vorlesungen über die Grundlegung der Mengenlehre, Wissensch. u. Hyp. Bd XXXI. J. v. Neumann, Die Axiomatisierung der Mengenlehre, Math. Zeitschr. 27, 1928.

4) I.e., more precisely, there exist undecidable propositions in which there occur no concepts beyond the logical constants $^{-}$ (not), $\vee$ (or), (x) (for all), $=$, except + (addition) $\times$ (multiplication), both in relation to natural numbers, and in which even the prefixes (x) are allowed to relate only to natural numbers. In such propositions, there can thus occur only numerical variables, but never function variables whether free or bound.

situation that lies in no way in, say, the special nature of the systems put up, but holds instead for a very wide class of formal systems to which belong especially all those that arise from both of the presented ones through the addition of finitely many axioms[5]), with the condition that no false propositions of the kind given in footnote[4]) become provable through the added axioms.

We shall sketch to begin with, before we go into the details, the main idea of the proof, naturally without raising any pretence to exactness. The formulas of a formal system (we delimit ourselves here on the system PM) are,

- 3 -

externally considered, finite series of basic signs (variables, logical constants, and brackets and points of separation), and it is easy to make it precise *which* series of basic signs are meaningful formulas and which not[6]). Proofs are analogously, from a formal point of view, nothing but finite series of formulas (with specific properties that can be given). It is, for metamathematical considerations, obviously indifferent what objects one takes as basic signs, and we decide to use natural numbers[7]) as such signs. A formula is then, correspondingly, a finite sequence of natural numbers[8]) and a proof figure a finite sequence of finite sequences of natural numbers. The metamathematical concepts (propositions) become hereby concepts (propositions) about natural numbers and sequences of such and therefore (at least in part) expressible within the system PM itself. One can show, especially, that the concepts "formula," "proof figure," "provable formula" are definable within the system PM, i.e., one can, for example, give a formula $F(x)$ of one free variable of the PM[9]) such that $F(x)$ states, interpreted contentfully: x is a provable

- 4 -

formula. We produce now an undecidable proposition of the system PM, i.e., a proposition A for which neither A nor non-A is provable, as follows:

---

[5]) In this, only those axioms are counted as distinct in PM that do not come out one from the other by a mere change of types.

[7]) I.e., we map the basic signs in a one-to-one way on the natural numbers.

[8]) I.e., a function over segments of the natural number sequence of natural numbers. [Belegung eines Abschnittes der Zahlenreihe mit natürliche Zahlen]

[9]) It would be very easy (just a bit long-winded) to actually write down this formula.

We shall call a formula from PM with one free variable of the type of the natural numbers (class of classes) a class sign. We think of the class signs as ordered in some way in a sequence and designate the n-th by $R(n)$ and note that the concept of "class sign" as well as the ordering relation R let themselves be defined in system PM. We designate by $[\alpha; n]$ the formula that arises from the class sign α through the replacement of the free variable by the sign for the natural number n[10]. Even the relation $x = [y, z]$ turns out to be definable within PM. We define now a class K of natural numbers as follows: $K(n) \equiv \overline{Bew}[R(n); n]$ [11]) (1)
(where $Bew(x)$ means: x is a provable formula). All of the concepts that occur in the *Definiens* are definable in PM, therefore also the concept K composed of them., i.e., there exists a class sign S [12]) such that

- 5 -

formula $[S; n]$ states, meant in a contentful way, that the natural number n belongs to K. S is, as a class sign, identical to a determinate $R(q)$, i.e., we have

$$S = R(q)$$

for a determinate natural number q. We show now that the proposition $[R(q), q]$ [13]) is undecidable in PM. For if it is assumed that the proposition $[R(q), q]$ is provable, then it would be correct contentfully interpreted, i.e., by the above, q would belong to K, i.e., by (1), $\overline{Bew}[R(q), q]$ would hold in contradiction with the assumption. If instead the negation of $[R(q), q]$ were provable, then $\overline{K(q)}$, i.e., $Bew[R(q), q]$ would hold. $[R(q), q]$ together with its negation would be provable, which is again impossible.

The analogy of this inference with the antinomy of Richard hits the eye; There is also a close relation with the "liar"[14]), for the undecidable proposition $[R(q), q])$ states that q belongs to K, i.e., by (1), that $[R(q), q]$ is not

[10]) In case α is not a class sign or n no natural number, one means by $[\alpha; n]$ the empty sequence of numbers, say.

[11]) Negation is denoted by overlining.

[12]) There is, again, not the least difficulty actually to write down the formula S.

[13]) As soon as S has been established, even q lets itself naturally be determined and thereby the undecidable proposition effectively written down.

[14]) In general, each epistemological antinomy can be turned into such a proof of undecidability.

provable. So we have a proposition in front of us that claims its own unprovability[15]).

-6-

The proof method just presented can be evidently applied to every formal system that, first, in terms of content, provides sufficiently in the form of means of expression, so that the concepts used in the above considerations (especially the concept "provable formula") can be defined, and, secondly, in which each provable formula is even contentfully correct. The exact carrying through of the above proof that is to follow has as its tasks among others the replacement of the second of the conditions presented by a purely formal and much weaker one.

2.

We go now into the exact carrying through of the proof sketched above and give, to start with, a precise description of the formal system P for which we want to prove the existence of undecidable propositions. P is essentially the system one obtains if one builds, upon the Peano axioms,

-7-

the logic of the *Princ. Math.*[16]) (numbers as individuals, successor relation as undefined basic concept).

The basic signs of the system are as follows:

I Constants: "$\sim$" (not), "$\vee$" (or), "Π" (for all, with the usage $x\Pi\varphi(x)$), "0" (zero), "f" (the successor of), "(", ")" (bracket symbols).

II Variables of the first type (for individuals, i.e., natural numbers): "x_1", "y_1", "z_1", ...

Variables of the second type (for classes of natural numbers): "x_2", "y_2", "z_2", ...

[15]) Such a proposition has, contrary to appearance, nothing circular about it, for it claims in the first place the undecidability of quite a specific formula (namely the q-th in the lexicographical ordering under a determinate substitution), and it turns out only afterwards (by chance as it were) that this formula is just the one in which the formula itself got expressed.

[16]) The addition of the axioms of Peano as well as all other changes called for in the system of the *Princ. Math.* serve merely for the simplification of the proof and can in principle be dispensed with.

Variables of the third type (for classes of classes of natural numbers): "x_3", "y_3" ,"z_3",...

etc for each natural number as a type[17]).

Variables for two-place functions (relations) are superfluous as basic signs, because one can define relations as classes of ordered pairs, and ordered pairs in turn as classes of classes, for example the ordered pair (a, b) by $((a), (a, b))$ in which (x, y) and (x) denote the classes the only elements of which are x, y and x, respectively[18]). We mean by a *sign of the first type* a combination of signs of the

- 8 -

form:

$$a \quad fa \quad ffa \ldots \overbrace{f \ldots fa}^{n} \ldots \qquad \text{etc}$$

in which a is either "0" or a variable of first type. In the first case, we call such a sign a *number sign*. For $n > 1$, we mean by a *sign of the n-th type* the same as a *variable of n-th type*. We call combinations of signs of the form $a(b)$, in which b is a sign of type n and a a sign of type $n + 1$, *elementary formulas*. We define the class of *formulas* as the smallest class to which belong the elementary formulas and to which belong, with a and b, always also $\sim (a)$, $(a) \vee (b)$, $x\Pi(a)$ (here x is an arbitrary variable[19]). We call $(a) \vee (b)$ the *disjunction* of a and b, $\sim (a)$ the *negation*, and $x\Pi(a)$ a *generalization* of a. A formula in which there occur no free variables (*free variable* defined in the known way) is called a *propositional formula.* We call a formula with exactly n free individual variables (and with no further free variables) an *n-place relation sign*, for $n = 1$ also *class sign.*

- 9 -

We mean by *Subst* $a\binom{v}{b}$ (in which a denotes a formula, v a variable, and b a sign of the same type as v) the formula that arises from a when v is replaced in it by b everywhere where it is free[20]). We say that a formula a is a *type*

17) It is required that denumerably many signs are available for each type of variable.

18) Even inhomogeneous relations can be defined in this way, for example, a relation between individuals and classes as a class with elements of the form: $((x_2), ((x_1, x_2))$ – A simple consideration shows that all theorems about relations provable in the *Princ. Math.* are provable also under this treatment.

19) So $x\Pi(a)$ is a formula even in the case that x does not occur or does not occur free in a. $x\Pi(a)$ means in this case naturally the same as a.

20) In case v is not free in a, *Subst* $a\binom{v}{b} = a$ shall be the case.

elevation of another b if a arises from b through the elevation by the same number of all the free variables that occur in b.

The following formulas (I to V) are called axioms (they are written with the help of the abbreviations defined in the usual way: $.$, $\supset$, $\equiv$, (Ex), $=$,[21]) and with the application of the usual conventions about the leaving out of brackets)[22]):

I 1. $\sim (fx_1 = 0)$

2. $fx_1 = fy_1 \supset x_1 = y_1$

3. $x_2(0) . x_1\Pi(x_2(x_1) \supset x_2(fx_1)) \supset x_1\Pi(x_2(x_1))$

II Each formula that arises from the following schemes through the substitution of arbitrary formulas for p, q, r.

1. $p \vee p \supset p$
3. $p \vee q \supset q \vee p$
2. $p \supset p \vee q$
4. $(p \supset q) \supset r \vee p \supset r \vee q)$

- 10 -

III Each formula that arises from the two schemes

1. $v\Pi a \supset Subst\, a\binom{v}{b}$

2. $v\Pi(b \vee a) \supset b \vee v\Pi(a)$

through making the following substitutions for a, v, b, c:

For a an arbitrary formula, for v an arbitrary variable, for b a formula in which v does not occur free, for c a sign of the same type as v, on the condition that c contain no variable that is bound in a in a place in which v is free[23]).

IV Each formula that arises from the scheme

1. $(Eu)(v\Pi[u(v) \equiv a])$

through the substitution of arbitrary variables of the types n and $n+1$ for

---

[21]) $x_1 = y_1$ is, as in the *Princ. Math.* I, $*13$, to be thought as defined by $x_2\Pi(x_2(x_1) \equiv x_2(y_1))$ (and the same for higher types).

[22]) To obtain the axioms from the schemes written down here one has, then (possibly after the execution of the allowed substitutions), to

1.) resolve the abbreviations, 2.) add the suppressed brackets.

[23]) So c is either a variable or a sign of the form $\underbrace{ff\ldots f}_{k}u$ in which u is either "0" or a numerical variable.

v and u, and for a a formula that does not contain u free. This axiom represents the axiom of reducibility (comprehension axiom in set theory).

V Each formula that arises from the following one through type elevation (and this formula itself):

$$x_1\Pi(x_2(x_1) \equiv y_2(x_1)) \supset x_2 = y_2$$

This axiom states that a class is completely determined by its members.

- 11 -

A formula c is called an *immediate consequence* of a and b (or of a) if a is the formula $(\sim(b)) \vee (c)$ (or if c is, respectively, the formula $v\Pi(a)$, where v is an arbitrary variable). The class of *provable formulas* is defined as the smallest class of formulas that contains the axioms and is closed with respect to the relation of "immediate consequence"[24]).

We associate next to the basic signs of system P natural numbers, in the following one-to-one way:

"0"...1 "∨"...7 "("...11

"f"...3 "Π"...9 ")"...13

"∼"...5

further, for the variables of type n the numbers of the form p^n (in which p is a prime number > 13). Hereby there corresponds to each finite series of basic signs (therefore also each to formula) in a one-to-one way a finite series of natural numbers. We map now the finite series of natural numbers (again in a one-to-one way) on the natural numbers, by letting the number $2^{n_1}.3^{n_2}\dots p_k{}^{n_k}$ correspond to the series $n_1, n_2 \dots n_k$, where p_k denotes the k-th prime number in size. Hereby a natural number is associated in a one-to-one way, not just to each basic sign, but also to each finite series

- 12 -

of such in a one-to-one way. We designate the number associated to the basic sign (or the finite series of basic signs) a by $\Phi(a)$. Next, let there be given whatever class or relation $R(a_1, a_2 \dots a_n)$ between basic signs or series thereof. We associate to it that class (relation) $R'(x_1, x_2 \dots x_n)$ between natural

24) The rule of substitution becomes superfluous through all possible substitutions having been carried out already in the axioms (analogously in J. v. Neumann, Zur Hilbertschen Beweistheorie, Math. Zeitschr 26, 1927).

numbers that obtains between $x_1, x_2 \ldots x_n$ when and only when there exist $a_1, a_2 \ldots a_n$ such that $x_i = \Phi(a_i)$ $(i = 1, 2, \ldots n)$ and $R(a_1, a_2 \ldots a_n)$ holds. We designate the classes and relations between natural numbers that are associated in this way to the metamathematical concepts so far defined, for example "variable," "formula," "sentence-sign," "axiom," "provable formula," etc, by the same words in cursive writing. The proposition that there exist formally unsolvable problems in system P, for example, reads as follows: There exists a *sentence formulas a* such that neither *a* nor the *negation* of *a* are *provable* formulas.

We put next up the following definition: A number-theoretic function $\varphi(x_1, x_2 \ldots x_n)$ [25] is said to be *recursively defined from* the number-theoretic functions $\psi(x_1, x_2 \ldots x_{n-1})$ and $\mu(x_1, x_2 \ldots x_{n+1})$ if the following holds for all $x_2 \ldots x_n, k$ [26]:

- 13 -

$$\begin{aligned} \varphi(0, x_2 \ldots x_n) &= \psi(x_2 \ldots x_n) \\ \varphi(k+1, x_2 \ldots x_n) &= \mu(k, \varphi(k, x_2 \ldots x_n), x_2 \ldots x_n) \end{aligned} \tag{2}$$

A number-theoretic function φ is said to be *recursive* if there exists a finite series of functions $\varphi_1, \varphi_2 \ldots \varphi_n$ that ends with φ and has the property that each of the functions φ_k of the series is either recursively defined from two preceding ones or arises from whatever of the preceding ones through substitution[27] or is finally a constant or identical to the successor function $x + 1$. The length of the shortest series of φ_i that belongs to a recursive function φ is called its *level* [Stufe]. A relation between natural numbers $R(x_1 \ldots x_n)$ is called recursive[28] if there is a recursive function $\varphi(x_1 \ldots x_n)$ such that for all $x_1, x_2 \ldots x_n$

$$R(x_1 \ldots x_n) \equiv [\varphi(x_1 \ldots x_n) = 0]\ ^{29)}$$

[25]) I.e., its domain of definition is the class of non-negative entire numbers and its range of values a proper or improper subclass thereof.

[26]) Lower case Latin letters (possibly with indices) are in the following always variables for non-negative entire numbers (in case the opposite is not especially noted).

[27]) More precisely, through substitution of some of the preceding functions in the argument places of one that precedes, for example:
$\varphi_k(x_1, x_2) = \varphi_p[\varphi_q(x_1, x_2), \varphi_r(x_2)]$ $(p, q, r < k)$.

[28]) Recursive relations R have clearly the property that one can decide for each specific n-tuple of numbers whether or not $R(x_1 \ldots x_n)$ holds.

[29]) For all considerations that relate to content, the Hilbertian symbolism is used.

The following theorems hold:

I *Each function (relation) that arises from recursive functions (relations) through substitution of recursive functions in the place of variables is recursive; the same for each function that*

- 14 -

arises from recursive functions by recursive definition after scheme (2).

II *If R and S are recursive relations, then also* $\overline{R}, R \vee S, R \& S$.

III *If the functions* $\varphi(\mathfrak{x})$ *and* $\psi(\mathfrak{y})$ *are recursive, then also the relation* $\varphi(\mathfrak{x}) = \psi(\mathfrak{y})$ [30]).

IV *If the function* $\varphi(\mathfrak{x})$ *and the relation* $R(x, \mathfrak{y})$ *are recursive, then also the relations* S, T

$$S(\mathfrak{x}, \mathfrak{y}) \sim (Ex)[x \leqq \varphi(\mathfrak{x}) \,\&\, R(x, \mathfrak{y})]$$
$$T(\mathfrak{x}, \mathfrak{y}) \sim (x)[x \leqq \varphi(\mathfrak{x}) \rightarrow R(x, \mathfrak{y})]$$

as well as the function ψ

$$\psi(\mathfrak{x}\mathfrak{y}) = \varepsilon_x[x \leqq \varphi(\mathfrak{x}) \,\&\, R(x, \mathfrak{y})],$$

where $\varepsilon_x F(x)$ means: the smallest x for which $F(x)$ holds, or 0 in case there is no such number x.

Theorem I follows immediately from the definition of "recursive." Theorems II and III depend, as one can easily convince oneself, on the recursiveness of the number-theoretic functions

$\alpha(x), \beta(x, y), \gamma(x, y)$

that correspond to the logical concepts $\neg, \vee, =$, namely:

$\alpha(0) = 1; \quad \alpha(x) = 0$ for $x \neq 0$

$\beta(0, x) = \beta(x, 0) = 0; \beta(x, y) = 1$, if x, y are both $\neq 0$

$\gamma(x, y) = 0$ if $x = y$; $\gamma(x, y) = 1$, if $x \neq y$

30) We use German letters $\mathfrak{x}\,\mathfrak{y}$ as abbreviating designations for arbitrary n-tuples of variables, e.g., $x_1\, x_2 \ldots x_n$.

- 15 -

The proof of theorem IV is in brevity the following: There is by assumption a recursive $\varrho(x, \mathfrak{y})$ such that: $R(x, \mathfrak{y}) \sim [\varrho(x, \mathfrak{y}) = 0]$. We define now by recursion scheme (2) a function $\chi(x, \mathfrak{y})$ in the following way:

$$\chi(0, \mathfrak{y}) = 0$$

$$\chi(n+1, \mathfrak{y}) = (n+1) \cdot a + \chi(n, \mathfrak{y}) \cdot \alpha(a)$$ [31)]

where $a = \alpha[\alpha(\varrho(0, \mathfrak{y}))] \cdot \alpha[\varrho(n+1, \mathfrak{y})] \cdot \alpha[\chi(n, \mathfrak{y})]$.

$\chi(n+1, \mathfrak{y})$ is therefore either $= n+1$ (when $a = 1$) or $= \chi(n, \mathfrak{y})$ (when $a = 0$) [32)]. The first case occurs clearly if and only if all of the factors of a are 1, i.e., when the following holds:

$$\overline{R}(0, \mathfrak{y}) \,\&\, R(n+1, \mathfrak{y}) \,\&\, [\chi(n, \mathfrak{y}) = 0]$$

From this it follows that the function $\chi(n, \mathfrak{y})$ (considered as a function of n) remains 0 until the smallest value n for which $R(n, \mathfrak{y})$ holds, and is from there on equal to this value (in case that $R(0, \mathfrak{y})$ already holds, we have correspondingly $\chi(n, \mathfrak{y})$ constant and = 0). By this we have:

$$\psi(\mathfrak{x}, \mathfrak{y}) = \chi(\varphi(\mathfrak{x}), \mathfrak{y})$$

$$S(\mathfrak{x}, \mathfrak{y}) \sim R[\psi(\mathfrak{x}, \mathfrak{y}), \mathfrak{y}]$$

- 16 -

The relation T can be reduced through negation back to a case that is analogous to S, by which theorem IV is proved.

The functions $x + y$, $x \cdot y$, x^y, further the relations $x < y$, $x = y$ are, as one is easily convinced, recursive, and starting with these concepts, we define now a series of functions (relations) 1–45, each of which is defined from the preceding ones by the procedures mentioned in theorems I to IV. Here, several of the steps of definition allowed by theorems I to IV are usually combined into one. Each of the functions (relations) 1–45, among them occur for example the concepts "FORMULA," "AXIOM," "IMMEDIATE CONSEQUENCE," is therefore recursive.

---

31) We assume as known that the functions $x + y$ (addition), $x \cdot y$ (multiplication) are recursive.

32) a cannot assume other values than 0 and 1, as is seen from the definition.

- 17 -

1. $x/y \equiv (Ez)[z \leqq x \,\&\, x = y \cdot z]$ [33]

   x is divisible by y. [34]

2. $Prim(x) \equiv \overline{(Ez)}[z \leqq x \,\&\, z \neq 1 \,\&\, z \neq x \,\&\, x/z] \,\&\, x > 1$

   x is a prime number.

3. $0 \, Pr \, x \equiv 0$

   $(n+1) Pr \, x \equiv \varepsilon_y [y \leqq x \,\&\, Prim(y) \,\&\, x/y \,\&\, y > n \, Pr \, x]$

   $n \, Pr \, x$ is the n-th (in order of value) prime number contained in x

4. $0! \equiv 1$

   $(n+1)! \equiv (n+1) \cdot n!$

5. $Pr(0) \equiv 0$

   $Pr(n+1) \equiv \varepsilon_y [y \leqq x\{Pr(n)\}! + 1 \,\&\, Prim(y) \,\&\, y > Pr(n)]$

   $Pr(n)$ is the n-th prime number (in order of value)

- 18 -

6. $n \, Gl \, x \equiv \varepsilon_y [y \leqq x \,\&\, x/(n \, Pr \, x)^y \,\&\, \overline{x/(n \, Pr \, x)^{y+1}}]$

   $n \, Gl \, x$ is the n-th member of the number series associated to x (for $n > 0$ and n not greater than the length of this series; otherwise 0)

7. $l(x) \equiv \varepsilon_y [y \leqq x \,\&\, y \, Pr \, x > 0 \,\&\, (y+1) \, Pr \, x = 0]$

   $l(x)$ is the length of the number series associated to x.

8. $x * y \equiv \varepsilon_z \{ z \leqq [Pr(l(x) + l(y))]^{x+y} \,\&$

   $(n)[n \leqq l(x) \rightarrow n \, Gl \, z = n \, Gl \, x] \,\&$

   $(n)[0 < n \leqq l(y) \rightarrow (n + l(x)) \, Gl \, z = n \, Gl \, y]\}$

   $x * y$ corresponds to the operation of "adjoining one to another" of two finite series of numbers.

---

33) The sign $\equiv$ is used in the sense of "definitional equality" (the symbolism is for the rest the Hilbertian one).

34) Whenever one of the symbols $(x), (Ex), \varepsilon_x$ occurs in the definitions to follow, it is followed by a bound on x. This bound serves merely to ensure the recursive nature of the concepts defined (cf. theorem IV). The *scope* of these concepts would, on the contrary, usually not change by the leaving out of this bound.

9. $R(x) \equiv 2^x$

$R(x)$ corresponds to the number series that consists of just the number x.

10. $E(x) \equiv R(11) * x * R(13)$

$E(x)$ corresponds to the operation of "bracketing" (11 and 13 are associated to the basic signs "(" and ")").

11. $n \, Var \, x \equiv (Ez)[13 < z \leqq x \,\&\, Prim(z) \,\&\, x = z^n] \,\&\, n \neq 0$

x is a VARIABLE OF TYPE n.

12. $Var(x) \equiv (En)[n \leqq x \,\&\, n \, Var \, x]$

x is a VARIABLE.

- 19 -

13. $Neg(x) \equiv R(5) * E(x)$

$Neg(x)$ is the NEGATION of x.

14. $x \, Dis \, y \equiv E(x) * R(7) * E(y)$

$x \, Dis \, y$ is the DISJUNCTION of x and y.

15. $x \, Gen \, y \equiv R(x) * R(9) * E(y)$

$x \, Gen \, y$ is the GENERALIZATION of y by the VARIABLE x.

16. $0 \, Nf \, x \equiv x$

$(n+1) Nf \, x \equiv R(3) * n \, Nf \, x$

$n \, Nf \, x$ corresponds to the operation: "setting the sign "f" n times in front of x."

17. $Z(n) \equiv n \, Nf[R(1)]$

$Z(n)$ is the NUMBER SIGN for the number n

18. $Typ_1(x) \equiv (Em, n)\{m, n \leqq x \,\&\, [m = 1 \vee 1 \, Var(m)]$

$\&\, x = n \, Nf[R(m)]\}$ [34a)]

x is a SIGN OF THE FIRST TYPE.

19. $Typ_n(x) \equiv [n = 1 \,\&\, Typ_1(x)] \vee [n > 1 \,\&$

[34a)] $m, n \leqq x$ stands for: $m \leqq x \,\&\, n \leqq x$ (and the same for more than 2 variables)

$$(Ev)\{v \leqq x \,\&\, n\, Var\, v \,\&\, x = R(v)\}]$$

x is a SIGN OF n-TH TYPE.

20. $Elf(x) \equiv (Ey,z,n)[y,z,n \leqq x \,\&\, Typ_n(y)$
$\&\, Typ_{n+1}(z) \,\&\, x = z * E(y)]$

x is an ELEMENTARY FORMULA.

- 20 -

21. $Op(xyz) \equiv x = Neg(y) \vee x = y\, Dis\, z \vee$
$(Ev)[v \leqq x \,\&\, Var(v) \,\&\, x = v\, Gen\, y]$

22. $FR(x) \equiv (n)\{0 < n \leqq l(x) \rightarrow Elf(n\, Gl\, x) \vee$
$(Ep,q)[0 < p,q \leqq n \,\&\, Op(n\, Gl\, x, p\, Gl\, x, q\, Gl\, x)]\}$
$\&\, l(x) > 0$

x is a series of FORMULAS each of which is either an ELEMENTARY FORMULA or comes out from the previous ones through the operations of NEGATION, DISJUNCTION, GENERALIZATION.

23. $Form(x) \equiv (En)\{n \leqq (Pr[l(x)]^2)^{x[l(x)]^2}$
$\&\, FR(n) \,\&\, x = [l(n)]\, Gl\, n\}$ [35)]

x is a FORMULA (i.e., the last member of a series of FORMULAS FR)

24. $v\, Fr\, n, x \equiv Var(v) \,\&\, Form(x) \,\&\, v = n\, Gl\, x \,\&$
$\overline{(Ea,b,c)}[a,b,c \leqq x \,\&\, x = a * (v\, Gen\, b) * c$
$\&\, Form(b) \,\&\, l(a) + 1 < n \leqq l(a) + l(v\, Gen\, b)]$

The VARIABLE v is free in x in the n-th place.

- 21 -

25. $v\, Geb\, n, x \equiv Var(v) \,\&\, v = n\, Gl\, x \,\&\, \overline{v\, Fr\, n, x}$

---

35) One recognises the bound $n \leqq (Pr[l(x)]^2)^{xl(x)^2}$ like this: the length n of the shortest series of formulas that belongs to x can be at most equal to the number of subformulas of x. But there are at most $l(x)$ subformulas of length 1, $l(x) - 1$ of length 2, and so on, so on the whole at most $\frac{l(x)[l(x)-1]}{2} < l(x)^2$. All of the prime numbers from n can then be assumed to be smaller than $Pr\{[l(x)]^2\}$, their number $< l(x)^2$, and their exponents (which are subformulas of x) $\leqq x$.

$\& \, Form(x)$

The VARIABLE v is BOUND at the n-th place in x.

26. $v \, Fr \, x \equiv (En)[n \leqq l(x) \,\&\, v \, Fr \, n, x]$

v occurs in x as a FREE VARIABLE.

27. $Su \, x\binom{n}{y} \equiv \varepsilon_z[(Eu, v) \, u, v \leqq x \,\&\, x = u * R(n \, Gl \, x) * v$
$\&\, z = u * y * v \,\&\, n = l(u) + 1]$

$Su \, x\binom{n}{y}$ arises from x when one substitutes y in place of the n-th member of x.

28. $0 \, St \, v, x \equiv \varepsilon_n\{n \leqq l(x) \,\&\, v \, Fr \, n, x$
$\&\, \overline{(Ep)}[n < p \leqq lx \,\&\, v \, Fr \, p, x]\}$

$(k+1) \, St \, v, x \equiv \varepsilon_n\{n < k \, St \, v, x \,\&\, v \, Fr \, n, x$
$\&\, \overline{(Ep)}[n < p < k \, St \, v, x \,\&\, v \, Fr \, p, x]\}$

$k \, St \, v, x$ is the $k+1$-th place in x (counted from the end of the FORMULA) in which v is FREE in x (and 0 in case there is no such place).

29. $A(v, x) \equiv \varepsilon_n\{n \leqq l(x) \,\&\, n \, St \, v, x = 0\}$

$A(v, x)$ is the number of positions in which v is FREE in x.

30. $Sb_0(x^v_y) \equiv x$

$Sb_{k+1}(x^v_y) \equiv Su[Sb_k(x^v_y)]\binom{k \, St \, v, x}{y}$

31. $Sb(x^v_y) \equiv Sb_{A(v,x)}(x^v_y)$ [36]

$Sb(x^v_y)$ is the concept $\text{Subst}(a^v_b)$ [37] defined above.

32. $x \, Imp \, y \equiv [Neg(x)] Dis \, y$

$x \, Con \, y \equiv Neg\{[Neg(x)] Dis [Neg(y)]\}$

$x \, Aeq \, y \equiv (x \, Imp \, y) \, Con (y \, Imp \, x)$

$v \, Ex \, y \equiv Neg\{v \, Gen [Neg(y)]\}$

33. $n \, Th \, x \equiv \varepsilon_y\{y \leqq x^{(x^n)} \,\&\, (k)[k \leqq l(x) \rightarrow$

[36]) In case v is not a VARIABLE or x not a FORMULA, we have $Sb(x^v_y) = x$.

[37]) We write, instead of $Sb[Sb(x^{\,v}_{\,y})^w_z]$, $Sb(x^{\,v\,w}_{\,y\,z})$ (analogously for more than two variables).

$$(k\,Gl\,x \leqq 13 \,\&\, k\,Gl\,y = k\,Gl\,x) \vee$$
$$(k\,Gl\,x > 13 \,\&\, k\,Gl\,y = k\,Gl\,x\,.\,[1\,Pr(k\,Gl\,x)]^n)]\}$$

$n\,Th\,x$ is the n-TH TYPE ELEVATION of x (in case x is a FORMULA).

Three specific numbers correspond to axioms I 1–3, ones we designate by z_1, z_2, z_3, and we define:

34. $Z\text{-}Ax(x) \equiv (x = z_1 \vee x = z_2 \vee x = z_3)$

- 23 -

35. $A_1\text{-}Ax(x) \equiv (Ey)[y \leqq x \,\&\, Form(y) \,\&\, x = (y\,Dis\,y)\,Imp\,y]$

x is a FORMULA that arises through a substitution in axiom II 1. $A_2\text{-}Ax, A_3\text{-}Ax, A_4\text{-}Ax$ that correspond to axioms II 2 to 4 are defined analogously.

36. $A\text{-}Ax(x) \equiv A_1\text{-}Ax(x) \vee A_2\text{-}Ax(x) \vee A_3\text{-}Ax(x) \vee A_4\text{-}Ax(x)$

x is a FORMULA that arises through a substitution in a propositional axiom.

37. $Q(z,y,v) \equiv \overline{(En,m,w)[n \leqq l(y) \,\&\, m \leqq l(z) \,\&\, w \leqq z \,\&}$
$$\overline{w = m\,Gl\,z \,\&\, w\,Geb\,n,y \,\&\, Fr\,n,y]}$$

z contains no VARIABLE that is BOUND in a place in y in which v is FREE.

38. $L_1\text{-}Ax(x) \equiv (Ev,y,z,n)\{v,y,z,n \leqq x \& n\,Var\,v \,\&$
$$Typ_n(z) \,\&\, Form(y) \,\&\, Q(z,y,v) \,\&$$
$$x = (v\,Gen\,y)Impl[Sb(y^v_z)]\}$$

x is a FORMULA that arises from axiom scheme III 1 through substitution.

39. $L_2\text{-}Ax(x) \equiv (Ev,q,r,p)\{v,q,r,p \leqq x \,\&\, Var(v) \,\&\, Form(p)$
$$\&\, \overline{v\,Fr\,p} \,\&\, Form(q) \,\&$$
$$x = [v\,Gen(p\,Dis\,q)]Impl[p\,Dis(v\,Gen\,q)]\}$$

x is a FORMULA that arises from axiom scheme III 2 through substitution.

- 24 -

40. $R\text{-}Ax(x) \equiv (Eu, v, y, n)[u, v, y, n \leqq x \,\&\, n\, Var\, v \,\&$
$(n+1)\, Var\, u \,\&\, \overline{u\, Fr\, y} \,\&\, Form(y) \,\&$
$x = u\, Ex\{v\, Gen[[R(u) * E(R(v))] Aeq\, y]\}]$

x is a FORMULA that arises from axiom scheme IV 1 through substitution.

A determinate number z_4 corresponds to axiom V 1 and we define:

41. $M\text{-}Ax(x) \equiv (En)[n \leqq x \,\&\, x = n\, Th\, z_4]$

42. $Ax(x) \equiv Z\text{-}Ax(x) \vee A\text{-}Ax(x) \vee L_1\text{-}Ax(x) \vee L_2\text{-}Ax(x)$
$\vee R\text{-}Ax(x) \vee M\text{-}Ax(x)$

x is an AXIOM.

43. $Fl(xyz) \equiv y = z\, Impl\, x \vee$
$(Ev)[v \leqq x \,\&\, Var(v) \,\&\, x = v\, Gen\, y]$

x is an IMMEDIATE CONSEQUENCE of y and z.

44. $Bw(x) \equiv (n)\{0 < n \leqq l(x) \rightarrow Ax(n\, Gl\, x) \vee$
$(Ep, q)[p, q \leqq n \,\&\, Fl(n\, Gl\, x, p\, Gl\, x, q\, Gl\, x)]\} \,\&\, l(x) > 0$

x is a PROOF FIGURE (a finite sequence of FORMULAS, each of which is either an AXIOM or an IMMEDIATE CONSEQUENCE of two preceding ones).

- 25 -

45. $x\, B\, y \equiv Bw(x) \,\&\, [l(x)]\, Gl\, x = y]$ [last bracket to be deleted]

x is a PROOF of the FORMULA y

46. $Bew(x) \equiv (Ey) y\, B\, x$

x is a PROVABLE FORMULA. ($Bew(x)$ is the only one among the concepts 1–46 of which it cannot be claimed that it is recursive).

- 26 -

The fact that one can vaguely formulate as: Each recursive relation can be defined within the system P (meant in a contentful way) *without* recourse to any contentful meaning of the formulas of P, is expressed in an exact form by the following theorem:

Theorem V: *There exists for each recursive relation* $R(x_1 \ldots x_n)$ *an n-place* RELATION SIGN r *(with the* FREE VARIABLES $u_1, u_2 \ldots u_n$*) such that for all* $x_1 \ldots x_n$ *holds:*

$$R(x_1 \ldots x_n) \rightarrow Bew[Sb(r^{u_1 \ldots\ldots u_n}_{Z(x_1) \ldots Z(x_n)})] \tag{3}$$

$$\overline{R}(x_1 \ldots x_n) \rightarrow Bew[Neg\, Sb(r^{u_1 \ldots\ldots u_n}_{Z(x_1) \ldots Z(x_n)})] \text{ }^{38)} \tag{4}$$

We rest content here to indicate the proof of this theorem in outline, as there are no difficulties of principle in it and as it is rather long-winded[39]). We prove the theorem for all relations $R(x_1 \ldots x_n)$ of the form:

$x_1 = \varphi(x_2 \ldots x_n)$[40])

(here φ is a recursive function) and apply complete induction on the level of φ. So let φ have level n. It arises from functions of a lower level $\varphi_1 \ldots \varphi_k$ through the operations of substitution or of recursive definition. By the inductive hypothesis, all is proved for $\varphi_1 \ldots \varphi_k$,

- 27 -

so there exist the associated RELATION SIGNS $r_1 \ldots r_k$ such that V holds. The definitional processes by which φ arises from $\varphi_1 \ldots \varphi_k$ (substitution and recursive definition) can all be formally reproduced in system P. If one does that, one obtains from $r_1 \ldots r_k$ a new RELATION SIGN r[41]) for which one can prove without difficulty that theorem V holds, under the condition that it holds for $\varphi_1 \ldots \varphi_k, r_1 \ldots r_k$. A RELATION SIGN r to which a recursive relation is associated in this way[42]) shall be called recursive.

We come now to the goal of our explanations: Let κ be an arbitrary class of FORMULAS. We designate by $Flg(\kappa)$ (set of consequences of κ [Folgerungsmenge]) the smallest set of FORMULAS that contains all FORMULAS of κ and all AXIOMS and is closed with respect to the relation of "IMMEDIATE

---

38) The VARIABLES $u_1 \ldots u_n$ can be given arbitrarily. There exists, for example, always an r with the FREE VARIABLES 17, 19, 23 ... etc for which (3) and (4) hold.

39) Theorem V depends obviously on the fact that with a recursive relation R it is, for each n-tuple of numbers, decidable *from the axioms of the system* P whether relation R obtains or not.

40) From this follows at once that it holds for each recursive relation, because such a relation is equal in meaning to $0 = \varphi(x_1 \ldots x_n)$, in which φ is recursive.

41) In a precise carrying through of this proof, r is not defined through the detour of a contentful meaning, but through a purely formal condition.

42) So, one that expresses, contentfully meant, that this relation obtains.

CONSEQUENCE." κ is called ω-consistent if there is no CLASS SIGN a (with the free variable v) such that:

$$(n)[Sb(a^{v}_{Z(n)})\,\varepsilon\,Flg(\kappa)]\ \&\ [Neg(v\,Gen\,a)]\,\varepsilon\,Flg(\kappa)$$

- 28 -

Each ω-consistent system is obviously also free from contradiction. The converse, however, does not hold, as will be shown later.

The general result about existence of undecidable propositions reads as follows:

Theorem VI: *There exists for each ω-consistent recursive class κ of* FORMULAS *a recursive* CLASS SIGN r *(with the* FREE VARIABLE v*) such that neither* $v\,Gen\,r$ *nor* $Neg\,v\,Gen\,r$ *belong to* $Flg(\kappa)$.

Proof: Let κ be an arbitrary recursive ω-consistent class of FORMULAS. We define:

$$Bw_{\kappa}(x) \equiv (n)[n \leqq l(x) \rightarrow Ax(n\,Gl\,x) \vee (n\,Gl\,x)\,\varepsilon\,\kappa \vee (Ep,q)\{p,q \leqq n\ \&\ Fl(n\,Gl\,x, p\,Gl\,x, q\,Gl\,x)\}]\ \&\ l(x) > 0 \quad (5)$$

(cf. the analogous concept 44)

$$x\,B_{\kappa}\,y \equiv Bw_{\kappa}(x)\ \&\ [l(x)]\,Gl\,x = y \quad (6)$$

$$Bew_{\kappa}(x) \equiv (Ey)\,y\,B_{\kappa}x \quad (6{\cdot}1)$$

(cf. the analogous concepts 45, 46)

We clearly have:

$$(x)Bew_{\kappa}(x) \sim x\,\varepsilon\,Flg(\kappa) \quad (7)$$

$$(x)Bew(x) \rightarrow Bew_{\kappa}(x) \quad (8)$$

- 29 -

We define now the relation:

$$Q(x,y) \equiv \overline{x\,B_{\kappa}[Sb(y\,^{19}_{Z(y)})]} \quad (8{\cdot}1)$$

Because $x\,B_{\kappa}\,n$ (by (6), (5)) and $Sb(y\,^{19}_{Z(y)})$ (by Def. 17, 31) are recursive, then also $Q(xy)$. Therefore there exists then, by theorem V and (8), a RELATION SIGN q (with the FREE VARIABLES 17, 19) such that the following hold:

$$\overline{x\,B_{\kappa}[Sb(y\,^{19}_{Z(y)})]} \rightarrow Bew_{\kappa}[Sb(q\,^{17}_{Z(x)}\,^{19}_{Z(y)})] \quad (9)$$

$$x\, B_\kappa[Sb(y\, {}^{19}_{Z(y)})] \rightarrow Bew_\kappa[Neg\, Sb(q\, {}^{17}_{Z(x)}\, {}^{19}_{Z(y)})] \tag{10}$$

We put:

$$p = 17Gen\, q \tag{11}$$

(p is a CLASS SIGN with the FREE VARIABLE 19)

and

$$r = Sb(q\, {}^{19}_{Z(y)}) \tag{12}$$

(r is a recursive CLASS SIGN with the FREE VARIABLE 17[43])). Then the following holds:

$$Sb(p\, {}^{19}_{Z(p)}) = Sb([17Gen\, q]\, {}^{19}_{Z(p)}) = 17Gen\, Sb(q\, {}^{19}_{Z(p)}) = 17Gen\, r \text{ [44])} \tag{13}$$

(because of (11) and (12)) further:

$$(x) Sb(q\, {}^{17}_{Z(x)}\, {}^{19}_{Z(p)}) = Sb(r\, {}^{17}_{Z(x)}) \tag{14}$$

\- 30 -

(by (12)). If one now substitutes in (9) and (10) p for y, then, taking into consideration (13) and (14), there arise:

$$\overline{x\, B_\kappa(17Gen\, r)} \rightarrow Bew_\kappa[Sb(r\, {}^{17}_{Z(x)})] \tag{15}$$

$$x\, B_\kappa(17Gen\, r) \rightarrow Bew_\kappa[Neg\, Sb(r\, {}^{17}_{Z(x)})] \tag{16}$$

Now there results:

1. $17Gen\, r$ is not κ-PROVABLE[45]). For were this the case, there would (by 6·1) exist an n such that $n\, B_\kappa(17Gen\, r)$. By (16), we would then have: $Bew_\kappa[Neg\, Sb(r\, {}^{17}_{Z(x)})]$, whereas on the other hand, from the κ-PROVABILITY of $17Genr\, r$ follows also that of $Sb(r\, {}^{17}_{Z(x)})$. So κ would be inconsistent (and the more so ω-inconsistent).

2. $Neg(17Genr)$ is not κ-PROVABLE, for, as was just proved, $17Gen\, r$ is not κ-provable, i.e., (by 6·1), we have $(n)\overline{n\, B_\kappa(17Gen\, r)}$. From this follows by (15) $(n) Bew_\kappa[Sb(r\, {}^{17}_{Z(x)})]$ that, together with $Bew_\kappa[Neg(17Gen\, r)]$, would

43) For r arises from the recursive relation sign q through the replacement of a free variable by a determinate number (p).

44) The operations Gen, Sb are obviously always exchangeable, in case they relate to different variables.

45) x is κ-PROVABLE shall mean $x\, \varepsilon\, Flg(\kappa)$ that says by (7) the same as $Bew_\kappa(x)$.

violate the ω-consistency of κ.

17*Gen r* is, then, undecidable from κ, by which theorem VI has been proved.

The proof of theorem VI was carried through without express consideration of intuitionistic requirements. It is, though, easy to convince oneself that the following is shown in an intuitionistically unobjectionable way: Let an arbitrary recursively defined class κ be presented. If, then, a formal decision (from κ) is laid out for the (effectively presentable) PROPOSITIONAL FORMULA 17 *Gen r*, one can effectively specify:

1. A PROOF for $Neg(17Genr)$.

2. A PROOF for $Sb(r\,^{17}_{Z(x)})$ for each arbitrary n, i.e., a formal decision of 17*Genr* would have as a consequence the effective demonstrability of an ω-contradiction.

- 30a -

We shall call a relation (class) $R(x_1 \dots x_n)$ between natural numbers *definite with respect to decision* if there is a RELATION SIGN r with the free variables $u_1 \dots u_n$ such that (3) and (4) hold. In particular, each recursive relation is then, by theorem V, definite with respect to decision. If a RELATION SIGN is associated in this way to a relation definite with respect to decision, it shall analogously be called *definite with respect to decision*. It is, now, sufficient for the existence of undecidable propositions to assume of the class κ that it is ω-consistent and definite with respect to decision. For definiteness with respect to decision is carried over from κ to $x\,B_\kappa\,y$ (cf. (6)) and to $Q(x,y)$ (cf. (8·1)), and only this was utilized in the above proof. The undecidable proposition has in this case the form v *Gen r*, in which r is a CLASS SIGN definite with respect to decision (it suffices, incidentally, that κ is definite with respect to decision in the system extended by κ).

If one requires of κ merely freedom from contradiction, instead of ω-consistency, there follows, even if not the existence of an undecidable proposition, then the existence of a property (r) for which neither a counterexample *can be given*, nor is it provable that it belongs to all numbers. For in the proof that 17 *Gen r* is not κ-PROVABLE, only the freedom from contradiction of κ was used, and from $\overline{Bew_\kappa}(17\,Gen\,r)$ follows by (15) $Sb(r\,^{17}_{Z(x)})$ for each number x, consequently for no number is $x\,Neg\,Sb(r\,^{17}_{Z(x)})$ κ-PROVABLE.

If one adjoins $Neg(17Gen\,r)$ to κ, one obtains a class of FORMULAS κ' that is free from contradiction but not ω-consistent. κ' is free from contradiction, because otherwise $17Gen\,r$ would be κ-provable. κ' is, though, not ω-consistent because we have by $\overline{Bew_\kappa}17Gen\,r$ and (15): $(x)Bew_\kappa Sb(r\,{}^{17}_{Z(x)})$, so even more also: $(x)Bew_{\kappa'}Sb(r\,{}^{17}_{Z(x)})$, and on the other hand there holds naturally: $Bew_{\kappa'}[Neg\,17Gen\,r)]$.[46)]

A special case of theorem VI is the one in which the class κ consists of finitely many FORMULAS (and those arising by TYPE ELEVATION). Each finite class α is obviously recursively definable. Let a be the greatest number contained in α. Then in this case the following holds for κ:

$$x\,\varepsilon\,\kappa \sim (Em,n)[m \leqq x \,\&\, n \leqq a \,\&\, n\,\varepsilon\,\alpha \,\&\, x = m\,Th\,n]$$

Then κ is recursive. This allows one to conclude, for example, that not all propositions are solvable even with the help of the axiom of choice (for all types) or with the generalized continuum hypothesis, assuming that these hypotheses are ω-consistent.

- 32 -

In the proof of theorem VI, no other properties of the system P were used than the following:

1.) The class of axioms and the rules of inference (i.e., the relation "immediate consequence") are recursively definable (as soon as one replaces in some way the basic signs by natural numbers).

2.) Each recursive relation can be defined inside system P (in the sense of theorem V).

Therefore there exist in each formal system in which the conditions 1.), 2.) are satisfied and that is ω-consistent undecidable propositions of the form $(x)F(x)$ in which F is a recursively defined property of natural numbers, and the same in each extension of such a system by a recursively defined class of axioms. To systems that satisfy the conditions 1.), 2.) belong, as is easy to ascertain, the axiom systems of set theory of Zermelo Fraenkel and of v. Neumann[47)], further the axiom system of number theory that

46) [The published version has here note 46. The typewritten manuscript jumps from note 45 to 47, see footnote 48 of the last shorthand version, p. 264L.)

47) The proof of condition 1 presents itself even as simpler than in the case of system P,

consists of the axioms of Peano, recursive definition (by scheme (2)), and the logical rules.[48]) Condition 1 is satisfied, in general, by each system the rules of inference of which are the usual ones and the axioms of which arise (analogously to P) through subsitution from finitely many schemes.

- 33 -

3

We derive now further consequences from theorem VI and give to this purpose the following definition:

A relation (class) is called arithmetic if it can be defined with the help of just the concepts $+, \cdot$ (addition and multiplication applied to natural numbers[49]) and the logical constants $\vee$, $\overline{}$, (x), $=$, in which (x) and = are allowed to apply only to natural numbers[50]). The concept of an "arithmetic proposition" is defined correspondingly. Especially, the relations "greater than" and "congruent modulo" for example are arithmetic, because we have:

$$x > y \sim \overline{(Ez)}[y = x + z]$$

$$x \equiv y\,(mod\, n) \sim (Ez)[x = y + z.n \vee y = x + z.n]$$

The following theorem holds:

Theorem VII: *Each recursive relation is arithmetic.*

We prove the theorem in the form: Each relation of the form $x_0 = \varphi(x_1 \ldots x_n)$ in which φ is recursive is arithmetic, and we use complete induction on the level of φ.

- 34 -

Let φ have level n. We have either:

1.) $\varphi(x_1 \ldots x_n) = \varrho[\chi_1(x_1 \ldots x_n), \chi_2(x_1 \ldots x_n) \ldots \chi_m(x_1 \ldots x_n)]$ [51])

---

because there is only one kind of ground variable (and two in v. Neumann).

[48]) Cf. problem III in D. Hilbert's talk on Probleme der Grundlegung der Mathematik, Math. Ann. 102.

[49]) 0 is counted here and in what follows always as belonging to the natural numbers.

[50]) The *Definiens* of such a concept must be built with only the help of the signs introduced and variables for natural numbers $x, y, \ldots$ (there must not occur function variables).

[51]) Obviously not all of the $x_1 \ldots x_n$ need in fact occur in the χ_i (cf. the example in footnote [27])).

(here ϱ and all of the χ_i have levels smaller than n) or:

2. $\varphi(0, x_2 \dots x_n) = \psi(x_2 \dots x_n)$

$\varphi(k+1, x_2 \dots x_n) = \mu[k, \varphi(k, x_2 \dots x_n), x_2 \dots x_n]$

(in which ψ, μ have a level lower than n).

In the first case, we have:

$$x_0 = \varphi(x_1 \dots x_n) \sim (Ey_1 \dots y_m)[R(x_0\, y_1 \dots y_m)\ \& \\ \& S_1(y_1, x_1 \dots x_n) \,\&\, \dots \,\&\, S_m(y_m, x_1 \dots x_n)],$$

in which R and S_i are the arithmetic relations that, by the inductive assumption, exist and are equivalent to $x_0 = \varrho(y_1 \dots y_m)$ and $y_i = \chi_i(x_1 \dots x_n)$, respectively. Therefore $x_0 = \varphi(x_1 \dots x_n)$ is in this case arithmetic.

In the second case, we apply the following procedure: One can define the relation $x_0 = \varphi(x_1 \dots x_n)$ with the help of the concept of a "sequence of numbers" (f) [52] in the following way:

$$x_0 = \varphi(x_1 \dots x_n) \sim (Ef)\{f_0 = \psi(x_2 \dots x_n) \,\&\, (k)[k < x_1 \rightarrow \\ f_{k+1} = \mu(k, f_k, x_2 \dots x_n)] \,\&\, x_0 = f_{x_1}\}$$

- 35 -

We have therefore, if $S(y, x_2 \dots x_m)$ and $T(z, x_1 \dots x_{n+1})$ are the arithmetic relations that, by the inductive assumption, exist and are equivalent to $y = \psi(x_2 \dots x_n)$ and $z = \mu(x_1 \dots x_{m+1})$, respectively:

$$x_0 = \varphi(x_1 \dots x_n) \sim (Ef)\{S(f_0, x_2 \dots x_n) \,\&\, (k)[k < x_1 \rightarrow \\ T(f_{k+1}, k, f_k, x_2 \dots x_n)] \,\&\, x_0 = f_{x_1}\} \tag{17}$$

We replace now the concept "sequence of numbers" by the concept "pair of numbers," by associating to the pair n, d the sequence of numbers $f^{(n,d)}$ $(f_k^{(n,d)} = [n]_{1+(k+1)d})$ in which $[n]_p$ denotes the smallest non-negative remainder of n modulo p.

We have then the

Lemma 1. If f is an arbitrary sequence of natural numbers and k an arbitrary natural number, there exists a pair of natural numbers n, d such that $f^{(n,d)}$ and f agree on the k first members.

[52] The $k+1$-st member of the sequence f is denoted by f_k (with f_0 the first).

Proof: Let l be the greatest of the numbers $k, f_0, f_1 \dots f_{k-1}$. Let n be determined so that:

$$n \equiv f_i (mod\, 1 + (i+1)l!) \quad \text{for } i = 0, 1, \dots k-1$$

something that is possible because each two of the numbers $1 + (i+1)l!$ $(i = 0, 1 \dots k-1)$ are relatively prime. For a prime number contained in two of these would also have to be contained in the difference

- 36 -

$(i_1 - i_2)l!$ and therefore, because of $i_1 - i_2 < l$, in $l!$, which is impossible. The pair of numbers [from the shorthand version: $n, l!$] fulfils therefore what is required.

The relation $x = [n]_p$ is defined by:

$$x \equiv n(mod\, p) \,\&\, x < n$$

It is therefore arithmetic, and then also the relation $P(x_0, x_1 \dots x_n)$:

$$P(x_0, \dots x_n) \equiv (En, d)\{S([n]_{d+1}, x_2 \dots x_n) \,\&\, (k)[k < x_1 \rightarrow T([n]_{1+d(k+2)}, k, [n]_{1+d(k+1)}, x_2 \dots x_n)] \,\&\, x_0 = [n]_{1+d(x_1+1)}\}$$

that is by (17) and Lemma 1 equivalent to: $x_0 = \varphi(x_1 \dots x_n)$ (with the sequence f in (17), the question is only about its course of values up to the x_1+1-st member.) Hereby theorem VII is proved.

In accordance with theorem VI, there exists for each problem of the form $(x)F(x)$ (F recursive) an equivalent arithmetic problem, and because the whole proof of theorem VII can be formalized within the system P, this equivalence is also provable. Therefore we have:

Theorem VIII: *There exist in each of the formal systems mentioned in theorem VI*[53]*) undecidable arithmetic propositions.*

- 37 -

The same holds (by the remark on page 32) for axiom system of set theory and their extensions by ω-consistent recursive classes of axioms.

We derive finally the following result:

Theorem IX: *There exist in all of the formal systems mentioned in theo-*

---

[53]) These are those ω-consistent systems that arise from P through the addition of a recursively defined class of axioms.

rem VI[53] *undecidable problems of the narrower functional calculus*[54]), i.e., formulas of the narrower functional calculus for which neither general validity nor the existence of a counterexample is provable[55]).

This depends on:

Theorem X: *Each problem of the form* $(x)F(x)$ *(F recursive) can be reduced back to the satisfiability of a formula of the narrower functional calculus.* (i.e., one can give for each recursive F a formula of the narrower functional calculus the satisfiability of which is equivalent to the correctness of $(x)F(x)$).

We count as formulas of the narrower functional calculus (e.F.) those formulas that are built up from the basic signs[56]) $^{-}, \vee, (x), =; x, y, \ldots$ (individual variables) $F(x), G(x,y), H(x,y,z)\ldots$ (class and relation variables), where (x) and =

- 38 -

are allowed to refer only to individuals. We join to these sings even a third kind of variables $\varphi(x), \psi(xy), \chi(xyz)$ etc that represent functions over objects (i.e., $\varphi(x), \psi(xy)$ etc denote unique functions the arguments and values of which are individuals[57]).) We shall call a formula that contains, beyond the basic signs of the e.F. introduced above, variables of the third kind ($\varphi(x), \psi(xy)\ldots$ etc) a formula in the extended sense (i.w.S.)[58]) [im weiteren Sinne]. The concepts "satisfiable," "generally valid" carry over without further into formulas i.w.S., and the theorem holds that gives, for a formula i.w.S. A, a usual formula B of the e.F. such that the satisfiability of A is equivalent to the satisfiability of B. B is obtained from A through

[54]) Cf. Hilbert-Ackermann, Grundzüge der theoretischen Logik.
By formulas of the narrower functional calculus in system P are to be understood those that arise from the formulas of the narrower functional calculus of the Princ. Math. through the replacement of relations by classes of higher type, as indicated on p. 7.

[55]) I have shown in my work: Über die Vollständigkeit der Axiome des logischen Funktionenkalküls, Monatsch. f. Math. u. Phys. XXXVII, 2, that each formula of the narrower functional calculus can be shown to be either generally valid, or there is a counterexample; the existence of this counterexample, though, is by theorem IX *not* always provable (in the formal system introduced).

[56]) Hilbert and Ackermann don't count the sign = among the narrower functional calculus. There exists, however, for each formula in which the sign = occurs one without this sign that is satisfiable when and only when the original one is (cf. the work cited in footnote 55).

[57]) Namely, the domain of definition shall always the *whole* of the domain of individuals.

[58]) Variables of the third kind are allowed to stand everywhere as placeholders for individual variables, for example $y = \varphi(x), F(x, \varphi(y)), G[\psi(x, \varphi(y)), x]$ etc.

replacing the variables $\varphi(x), \psi(xy)$ that occur in A by expressions of the form: $(\imath z)F(zx), (\imath z)G(z,xy)\ldots$, the descriptive functions in the sense of the Princ. Math. I∗14 resolved and the formula obtained in this way logically multiplied[59)] by an expression that states that all of the F put in place of φ are unique in respect of the first placeholder.

We show now that there is to each problem of the form

- 39 -

$(x)F(x)$ (F recursive) an equivalent one that concerns the satisfiability of a formula i.w.S. from which theorem X follows by the remark made.

Since F is recursive, there is a recursive function $\Phi(x)$ such that $F(x) \sim [\Phi(x) = 0]$, and there is for Φ a series of functions $\Phi_1, \Phi_2 \ldots \Phi_n$ such that $\Phi_n = \Phi, \Phi_1(x) = x+1$, and for each Φ_k $(1 < k \leqq n)$ either:

$$1.)\ (x_2 \ldots x_m)[\Phi_k(0, x_2 \ldots x_m) = \Phi_p(x_2 \ldots x_m)] \tag{18}$$

$$(x, x_1 \ldots x_m)\{\Phi_k[(\Phi_1(x), x_2 \ldots x_m] = \Phi_q[x, \Phi_k(x, x_2 \ldots x_m), x_2 \ldots x_m]$$

$$p, q < k$$

or:

$$2.)\ (x_1 \ldots x_m)[\Phi_k(x_1 \ldots x_m) = \Phi_r(\Phi_{i_1}(\mathfrak{x}_1) \ldots \Phi_{i_s}(\mathfrak{x}_s))]^{60)} \tag{19}$$

$$r < k,\ i_v < k \text{ (for } v = 1, 2 \ldots s)$$

or:

$$3.)\ (x_1 \ldots x_m)[\Phi_k(x_1 \ldots x_m) = \Phi_1(\Phi_1 \ldots \Phi_1(0))] \tag{20}$$

We build further the propositions:

$$(x)\overline{\Phi_1(x) = 0}\ \&\ (xy)[\Phi_1(x) = \Phi_1(y) \rightarrow x = y] \tag{21}$$

$$(x)[\Phi_n(x) = 0] \tag{22}$$

We replace now in all the formulas (18), (19), (20) (for $k = 2, 3 \ldots n$) and in (21) (22) the functions Φ_i by function variables φ_i, the number 0 by an individual variable x_0 that has no occurrences otherwise, and form the conjunction C of all the formulas obtained in this way.

59) I.e., the conjunction formed.

60) $\mathfrak{x}_i$ $(i = 1, \ldots s)$ represent arbitrary n-tuples of the variables $x_1, x_2 \ldots x_m$, for example: $x_1 x_3 x_2$.

- 40 -

The formula $(Ex_0)C$ has then the required property, i.e.

1. If $(x)[\Phi(x) = 0]$ holds, $(Ex_0)C$ is satisfiable, for the functions $\Phi_1, \Phi_2 \ldots \Phi_n$ clearly give, when substituted in $(Ex_0)C$ for $\varphi_1, \varphi_2 \ldots \varphi_n$, a correct proposition.

2. If $(Ex_0)C$ is satisfiable, $(x)[\Phi(x) = 0]$ holds.

Proof. Let $\Psi_1, \Psi_2 \ldots \Psi_n$ be the functions that exist by assumption and that deliver, when substituted in $(Ex_0)C$, a correct proposition. Let their domain of objects be $\mathfrak{J}$. Because of the correctness of $(Ex_0)C$ for the functions Ψ_i, there exists an individual a (from $\mathfrak{J}$) such that all of the formulas (18) to (22) turn into correct propositions (18′) to (22′) under the replacement of the Φ_i by Ψ_i and of 0 by a. We build now the smallest subclass of $\mathfrak{J}$ that contains a and that is closed with respect to the operation $\Psi_1(x)$. This subclass ($\mathfrak{J}'$) has the property that each of the functions Ψ_i, when applied to elements from $\mathfrak{J}'$, gives again an element of $\mathfrak{J}'$. For this holds for Ψ_1 by the definition of $\mathfrak{J}'$, and because of (18′), (19′), (20′), this property is carried over from Ψ_i with a lower index to ones with a higher one. We call Ψ_i' the functions that arise from the Ψ_i through a limitation to the domain of individuals $\mathfrak{J}'$. All of the formulas (18)–(22) hold for these functions (under the replacement of 0 by a and of Φ_i by Ψ_i').

- 41 -

The individuals from $\mathfrak{J}'$ can be, because of the correctness of (21) for Ψ_1' and a, mapped in a unique way to the natural numbers, and moreover so that a goes over to 0 and the function Ψ_1' to the successor function Φ_1. By this mapping, all the functions Ψ_i' go over to the functions Φ_i' and because of the correctness of (22) for Ψ_n' and a, $(x)[\Phi_n(x) = 0]$ or $(x)[\Phi(x) = 0]$ holds, as was to be proved[61]).

The considerations that have led to the proof of theorem IX (for each specific F) can be carried through also within the system P. Therefore the equivalence between a proposition of the form $(x)F(x)$ (F recursive) and the satisfiability of the corresponding formula of the e.F. is provable in P, and therefore the undecidability of one of the propositions follows from

[61]) It follows from theorem IX that, for example, the problems of Fermat and that of Goldbach would become solvable if one had solved the decision problem of the e.F.

that of the other, by which theorem X has been proved[62]).

To finish, let us point at the following interesting circumstance that concerns the undecidable proposition *A* put up in the above. By a remark made right in the beginning,

A claims its own unprovability. Because *A* is undecidable, it is naturally also unprovable. Then, what *A* claims is correct. We have, then, decided with the help of metamathematical considerations a proposition *A* that is undecidable in the system. An exact analysis of this state of affairs leads to interesting results that concern a proof of freedom from contradiction of the system P (and related systems) that will be treated in a continuation of this work soon to appear.

[62]) Theorem X holds obviously also for the axiom system of set theory and its extensions through recursively definable ω-consistent classes of axioms, because there exist even in these systems undecidable propositions of the form $(x)F(x)$ (F recursive).

Part V

Lectures and seminars on incompleteness

J. von Plato, *Can Mathematics Be Proved Consistent?*, Sources and Studies in the History of Mathematics and Physical Sciences, https://doi.org/10.1007/978-3-030-50876-0_5

Lecture on undecidable propositions (Bad Elster)

–1–

A formal system S is, as is well known, complete (definite with respect to decision) if each proposition p expressible in the symbols of S is decidable from the axioms of S, i.e., either p or non-p derivable from the axioms in finitely many steps by the rules of inference of the calculus of logic. In the following, a procedure shall be sketched that shows, not only that all of the formal systems so far put up for mathematics (*Principia Mathematica*, axioms systems of set theory, systems of the Hilbert school) are incomplete, but beyond this allows to prove quite generally: Each formal system of finitely many axioms that contains the arithmetic of natural numbers is incomplete. The same holds also for systems with infinitely many axioms, provided that the axiom rule (i.e., the law by which the infinite set of axioms is produced) is constructive (in a sense that can be made precise[1]). One can indicate, for each formal system that satisfies the conditions mentioned, effectively an undecidable proposition, and the

– 2 –

propositions thus constructed belong to the arithmetic of natural numbers. Here those concepts and propositions are counted as *arithmetic* that are expressible solely through the concepts of addition and multiplication and the logical connectives (not, or, and, all, there is). "All" and "there exists" may refer here only to natural numbers.

The proof procedure that delivers this result runs as follows: Let the formulas of the system at hand be first numbered (there are naturally only denumerably many), in an arbitrary way that is fixed once and for all. By this, there is associated to each concept that pertains to formulas (each metamathematical concept) a definite concept that pertains to natural numbers, e.g., to the metamathematical relation "formula a is derivable from formula b by the rules of inference" is associated the following relation R between natural numbers: R obtains between the numbers m and n if and only if

[1] One can think of the axiom rule as being given in the form of something like a law that associates to each natural number n an axiom. One can call it constructive in a comprehensive sense if the law provides a procedure that allows to write down effectively for each number n the axiom that belongs to it. The concept of "constructive" that lies at the basis of the above theorem is, admittedly, more strict in its wording, but no law is known that would be constructive in one of these senses but not in the other.

the formula with number m is derivable from the formula with number n. To the concept "provable formula" corresponds the class of those numbers that are numbers of provable formulas, etc. It turns out, now, (as a detailed investigation shows), that the relations between numbers (and classes of numbers) defined in such a way by a detour over metamathematics, differ in principle in no way from the relations and classes that otherwise occur in arithmetic (e.g., "prime number," "divisible," etc).

– 3 –

They can be reduced just like that back to such through definition, without taking first the detour over the metamathematical concepts; i.e., more precisely: these relations turn out to be arithmetic in the above sense. That this is the case depends in the end on the fact that the metamathematical concepts pertain solely to certain combinatorial relations between formulas that are directly reflected in the associated numbers (under a suitable association).

We shall consider now the totality of propositional functions in one variable that are contained in the formal system at hand, and think of these as ordered in a sequence.

$$\varphi_1(x), \varphi_2(x), \dots \varphi_n(x) \dots \tag{1}$$

Next we define a class K of natural numbers in the following way:

$$n \,\varepsilon\, K \equiv \sim Bew\, \varphi_n(n) \tag{2}$$

Here *Bew* x shall mean: x is a provable formula. The class K was defined through the detour over metamathematical concepts ("propositional function," "provable" etc.). This detour can be, as noted above, avoided, i.e., one can give an arithmetic class (in the above sense) that is coextensive with K. It is the salient point for what follows that this is in fact possible, and it must naturally be proved in detail, something to which we shall, however, not enter here. The condition was that arithmetic

– 4 –

is contained in the system S, so there is in S and therefore in series (1) a propositional function $\varphi_k(x)$ that is coextensive with K, for which we then have:

$$\varphi_k(n) \equiv \sim Bew\, \varphi_n(n) \tag{3}$$

If one puts k in place of n, it follows:

$$\varphi_k(k) \equiv \sim Bew\, \varphi_k(k) \tag{4}$$

From (4) results, however, that neither $\varphi_k(k)$ nor $\sim \varphi_k(k)$ can be provable, for from the assumption that $\varphi_k(k)$ is provable would follow that $\varphi_k(k)$ is correct, and therefore, because of (4), it would not be provable. From the assumption that $\sim \varphi_k(k)$ is provable follows that $\sim \varphi_k(k)$ is correct and therefore, because of (4), $\varphi_k(k)$ is provable that is as well in contradiction with the assumption. A tacit condition in the proof was that each proposition provable in system S be correct. This condition can be replaced by a much weaker one, as a closer investigation shows, one that requires just a little more than the consistency of the formal system considered.

The procedure just sketched delivers, for each system that satisfies the conditions mentioned, an arithmetic proposition undecidable in this system. This proposition is, though, by no means absolutely undecidable, one can instead go always over to "higher" systems in which the proposition in question becomes decidable (there remain obviously other undecidable propositions). It results, especially, that for example analysis is a higher system in this sense than number

– 5 –

theory, and the axiom system of set theory again higher than analysis. It follows, for example, that there exist number-theoretic problems that cannot be solved by number-theoretic means, but only with analytical or set-theoretic ones.

A result that concerns proofs of freedom from contradiction is revealed by the above investigations. The proposition that a system is free from contradiction is a metamathematical proposition. Therefore it can be replaced by an arithmetic one by the above procedure (therefore expressible in the same system). It turns out that this proposition is always unprovable in the system the freedom from contradiction it claims (under the same conditions as above). I.e., the freedom from contradiction of formal systems (of the kind characterised above) can never be shown by lesser (or the same) ways of inference than are formalized in the system in question, one needs instead for such always some ways of inference that go beyond the system.

On formally undecidable propositions (Bad Elster)

–1–

The development of research into foundations of the past decades allows many a question to have a precise formulation and a mathematical treatment, something that earlier had to be left over to subjective opinions. One such question is, e.g., the decidability of each precisely posed mathematical problem. To get to grips with this question (to give it a precise sense at all), an analysis of the proof methods used in mathematics was required above all. This analysis was delivered by the logicists (Frege etc) and it has revealed that all the ways of inference used in today's mathematics can be reduced back without residue to a few axioms and ways of inference. The creation of an exact formula language as a substitute for the imprecise word-language went hand in hand with this analysis.

Formal systems of this kind, i.e., formula languages with the axioms and rules of inference that belong to them, have been put up for different areas of mathematics (number theory etc), and the question could then be put precisely whether these systems are complete in the sense that each proposition of the discipline in question is decidable in it, i.e., the proposition itself or its negation provable from the axioms in a finite number of steps. With the higher disciplines, especially abstract set theory, there was anyway doubt whether this is the case, but with the arithmetic of the natural numbers for example, it was generally believed. The completeness of the formal system in question was even repeatedly expressed as a conjecture. What I would like to show you here is [cancelled: that a complete formalization is impossible already with the arithmetic of the natural numbers] the surprising fact that there exist undecidable propositions already in systems of number theory, and even more, namely that: In each formal system, however described, in which the arithmetic propositions

– 2 –

are expressible, there certainly exist undecidable arithmetic propositions, with the condition that results by itself that no false, i.e., contentfully contradictory arithmetic propositions be provable in the system in question.

I shall make next precise what is to be understood by a formal system. One speaks of a formal system when the following are given

1.) A certain finite or denumerably infinite set of basic signs.

2.) Precise rules for how the formulas are built up from these basic signs, i.e., which combinations of basic signs are formulas and which not.

3.) A certain finite set of formulas has to be singled out as axioms.

4.) Certain rules of inference, finitely many, have to be given which allow to derive new formulas from the axioms and formulas already proved.

A formula *A* is said to be provable in the system in question if there exists a finite series of formulas that begins with whichever of the axioms and ends with formula *A* and which has the further property that each formula of the series arises through application of a rule of inference from whichever previous ones. A series of formulas with this property is even called a proof figure.

I would like to remark further that the condition by which the axioms must be at hand in only a finite number is not required for the existence of undecidable propositions. Even when a formal system has infinitely many axioms, naturally denumerably many, in which connection there has to be given a rule that allows to actually

– 3 –

write down the *n*-th axiom for each *n*, the proof that follows is applicable, admittedly under a certain condition for this so-called axiom rule that, however, is *in praxi* always satisfied, i.e., one cannot give a single counter-example.

It is perhaps appropriate to describe as arithmetization of metamathematics the method by which this very general result is achieved. One understands by metamathematics, as is known, that discipline which is preoccupied with the study of formal systems as I have characterised them above. The object of metamathematics is formed, then, of formulas, i.e., combinations of signs and their properties and relations, in the same sense in which the object of geometry is formed of points, lines, planes, and their properties and relations (examples of metamathematical concepts). The decisive circumstance, now, is the following: Just as one can carry over the geometric concepts and propositions, through a one-to-one mapping of points on triples of numbers, into analytical concepts and propositions, so one can equally carry over the metamathematical concepts and propositions into arithmetic, through a one-to-one mapping of formulas on numbers. Here the question is naturally of a mapping on the natural numbers, because there are in each formal system only denumerably many formulas.

This arithmetization has, further, a special consequence in metamathematics. The metamathematical propositions go in this way over into arithmetic propositions and are, then, expressible in the symbolism of the formal system of mathematics. There follows the strange circumstance that the theory that has as its object a certain formal system, is even

– 4 –

expressible by the formulas of this system and this is, as you shall soon see, the decisive circumstance in the proof to follow. Namely, it will become possible in this way to formally reconstruct antinomies of the second kind (Richard in the usual sense), naturally in a way that they, divested of their character as antinomies, turn into correct proofs of certain metamathematical facts.

I shall, next, go over to the proof and assume that some formal system in the above sense is at hand. We consider the totality of the propositional functions and in particular the one-place functions that belong to this system. By a one-place propositional function one understands, as is known, a formula with one free variable that becomes a proposition as soon as one substitutes for this variable an object of a determinate type. We encounter in particular as such a type of objects the natural numbers, because arithmetic has, clearly, to be contained in the system under consideration.

I have written down here an example of such an arithmetic propositional function. The variable is n and the formula states that the great theorem of Fermat with the number n as an exponent is correct ...

Each such propositional function expresses, then, a certain property of numbers, and a specific class of numbers is associated to it, namely the class of those numbers that have the property, i.e., ones that when substituted give a correct proposition. (In our case ...)

There exist naturally only denumerably many one-place propositional functions in each formal system, because there exist on the whole only denumerably many formulas, and we think of all of these as ordered in a sequence.

– 5 –

We shall derive out of this sequence of propositional functions a sequence of propositions, through substituting always in the n-th propositional function the number n. This set of propositions decomposes, now, into two classes, namely those that are provable in the system under question and those that are not, and I designate by K that class of indices for which the associated proposition of the above series is not provable. We have, then, defined in this way a certain class K of natural numbers. The definition is written down here in formulas. This definition was made, as you can see, through the detour over metamathematical concepts. There appear, indeed, the concepts of propositional function and provable formula in it.

It is, now, possible to avoid the detour, by the arithmetization of metamathematics previously mentioned, and to define the class K directly in an arithmetic way. The arithmetic concepts, though, are expressible in the formal system laid as a basis (something that was required). One can then express the property of natural numbers, to belong to class K, in the symbols of the system in question, through a certain propositional function $\varphi(x)$ for which we then have: $\varphi(x)$ is equivalent to the n-th proposition of the above series not being provable. But $\varphi(x)$ must, as a propositional function of the system laid as a basis, occur in the above series, i.e., be identical with a determinate $\varphi_q(x)$. Then (5) holds[1] — and now I claim that the proposition that results when one substitutes q in this propositional function, that is to say the proposition $\varphi_q(q)$, is undecidable. For! We do have that if we substitute in formula (5) for n the number p, then (6),[2] or if we denote proposition $\varphi_q(q)$ abbreviated by P, then (7).[3] Proposition P is hence equivalent with its own unprovability.

– 6 –

It follows then at once that neither P nor $\sim P$ can be provable, if we take into account in addition that each provable proposition is also correct (something that was also presupposed).

I would also wish to note that the number q and thereby the undecidable proposition P can always be actually determined, i.e., one can write down effectively for each formal system an undecidable arithmetic proposition. It results from a closer analysis that the undecidable proposition thus constructed is of a relatively simple structure, i.e., more precisely, there occur in it beyond the logical constants no other concepts than addition and multiplication applied to natural numbers. In the proof that I have

[1] [By the short version, (5) is $\varphi_q(n) \equiv \sim Bew\ \varphi_n(n)$. The letter has been changed from p to q.]

[2] [With q instead of p, (6) is $\varphi_q(q) \equiv \sim Bew\ \varphi_q(q)$.]

[3] [Condition (7) is $P \equiv \sim Bew\ P$.]

just sketched, the condition was used essentially by which each proposition provable in the system is even contentfully correct, i.e., that no false propositions are provable. It is a condition the correctness of which could even be doubted, but it can be replaced by a purely formal one that relates to no contentful meaning (correctness) of the formulas. This formal property of the system that one must presuppose is indeed, in a certain sense, a sharpening of the property of freedom from contradiction, into which I, however, cannot go closer here.

A strange consequence arises from the above result, regarding the question of freedom from contradiction, to which I shall now move. The question at hand is the following one. The proposition by which a certain formal system is free from contradiction is a metamathematical proposition. It states that a certain formula cannot occur as the endformula of a proof figure.

– 7 –

The arithmetization of metamathematics turns therefore the statement of freedom from contradiction into a certain arithmetic proposition that can under the circumstances be expressed in the symbols of the self-same formal system the freedom from contradiction of which it asserts, and this result reads now that this arithmetic proposition is always unprovable in the system under consideration, i.e., the freedom from contradiction of a formal system is never provable by the methods of proof that are formalizable in this system.

The train of thought in the proof is the following: Arithmetic statements of a certain constitution have the property that if they are correct, they are then for sure even provable from the arithmetic axioms, e.g., the negation of Goldbach's conjecture is such a statement. If it is correct that certain even numbers are not a sum of two prime numbers, then that is certainly also provable, for it is in that case possible to write down the even number in question and to determine by a finite number of trials that it is not the sum of two primes. This consideration does not hold, for example, for the Goldbach conjecture itself. It would be without further ado possible that each single even number is the sum of two prime numbers without there existing a proof for it from the arithmetic axioms.

A closer investigation shows that even the negation of the undecidable proposition has the mentioned property, namely that one can infer from its correctness to its provability. Then the above formula (8) holds.[4] For let us assume that $\sim P$ held. Then $\sim P$ would be provable, on the basis of (8). On the basis of (7) even P would also be provable, i.e., it follows from $\sim P$ that

– 8 –

P as well as $\sim P$ would be provable, i.e., that the formal system in question would be contradictory, or the other way around, P follows from the freedom from contradiction of the system as expressed by formula (9).[5]

One can carry through the proof for formula (9) as I have sketched here, also purely formally within arithmetic, and from this follows: Were the freedom from contradiction provable, then also P provable, whereas it was shown earlier that P is undecidable, consequently also unprovable. The freedom from contradiction of the formal system considered is then not provable in this system

This result, the proof of which I have naturally sketched to you only in quite rough lines, has for example as a consequence that the freedom from contradiction of classical mathematics is not provable by even using all of the set-theoretic and analytical ways of inference of classical mathematics, even less then by a proof apparatus somehow restricted. The formalistic school searches for a proof of the freedom from contradiction of classical mathematics by finite means, i.e., there must occur in the proof only decidable properties and computable functions, and what is called the existential way of inference must not be applied anywhere. But all finite ways of inference are easily formalizable in the system of classical mathematics and it is not at all foreseeable today how one could find ones that are not formalizable, even if one cannot exclude this with absolute certainty.

– 9 –

To end with, I would like to make notice that in all the considerations so far, the question has always been of relative undecidability and unprovability, i.e., of undecidability and unprovability in specific formal systems given at hand. One can extend these systems through the addition of new axioms in a way in which the undecidable propositions become solvable, indeed, this extension results in an entirely natural and so to say compelling way, through the introduction of certain ever higher concept formations, i.e., in the language of logicism, higher types and the axioms that

[4] [This should be $\sim P \equiv Bew \sim P$.]

[5] [This should be $\sim (Bew\, P \,\&\, Bew \sim P) \supset P$.]

belong to them. The decisive thing is, though, that however far one should go in this extension of concept formation, there remain always undecidable propositions, because one can always apply the proof methods that I have presented to you also to the extended systems. It is, then, certain that one will never come to a system in which all arithmetic propositions would be decidable.

[The reverse of page 9 has the following written in ink:]

1.) Zermelo

2.) [Cancelled: As a clarification to your remark] end of my talk work I believe I have explained sufficiently see also what I mean by a formal system

4.) [4 written over 3, then cancelled: Clarification] Zermelo talk delivered

3′ Skolemism

[The bottom of the folded page has the text: Über formal unentscheidbare Sätze, frühere Fassung]

867[1]

I On undecidable propositions (Vienna)

As soon as a domain of mathematics (say, geometry) is axiomatized, the question of the completeness of this axiomatization presents itself. One can understand different things with this. The meaning that is closest to what I lay here as a foundation is the following. An axiom system shall be called complete if each appropriate proposition (i.e., each proposition that deals with the basic concepts that occur in the system) can be decided by logical inferences from the axioms, i.e., whenever always either the proposition itself or its negation can be logically inferred from the axioms.

For such a question (about the completeness of an axiom system) to have a precise sense, one must require that the term "logical inference" is made precise in the first place. Namely, this one is by no means as unequivocal and clear as it might appear on a first sight. Let me just remind of the axiom of choice that is seen as logically precise by quite a few mathematicians, by others not, and of the law of excluded middle that was shown, as is known, dubitable by Brouwer. Finally, one should consider that careless uses of the "substitution" inference can even lead to a contradiction (to the antinomies of set theory).

For the question of completeness to have a precise sense at all, it is absolutely required that the logical ways of inference allowed to be used in derivations from an axiom system be made precise. This kind of making precise is achieved through what is known as *formalization* of the discipline in question. Formalization goes, then, beyond axiomatization, in that even the logical inferences are axiomatized in it, whereas one assumes such to be given by nature, so to say, in axiomatics. To achieve this goal of an axiomatic conception of logical inference, the imprecise and often equivocal word-language [Wortsprache] needs to be replaced by an exact formula language to start with. It allows the expression of each proposition of the discipline in question in an unequivocal way through a formula (inexact language is inexact logic). Secondly, one has to adjoin to the simple formulas that ex-

[1] [These clearly written pages are preceded by a page with the longhand heading "Formeln" and a whole page of symbolic expressions. The first ones can be placed within the text where some empty space is left for them, the latter part is similar to those found on a leaf of formulas midway through the lecture. I have placed them there, in continuation of the others.]

press the axioms of the discipline in question certain others that express the logical axioms (say, the law of contradiction, etc). Third, one must list *in concreto* the rules of inference by which it shall be allowed to derive further formulas from the axioms. These rules of inference must be formulated so that they refer only to the form of the formulas (not their meaning). Expressed in another way, it must be possible to use these rules of inference purely mechanically, even for someone who doesn't know the meaning of the formulas.[2]

With these, we have agreed on the essential characteristics of a formal system.

A formal system lies, then, at hand when the following are given:

1. Certain basic signs (the undefined concepts of the system) in a finite number.

2. It has to be made precise how the meaningful propositions of the system are built from these basic signs. i.e., it has to be specified which combinations of the basic signs are meaningful formulas and which not. It is the case even in a natural language

868

that not just any combination of words expresses a proposition. (Here one must guard oneself all the time against mixing the false and senseless. The formula $1 = 1 + 1$, for example, is definitely a meaningful proposition. It claims something but it is wrong. The combination of signs $= +1$, instead, is meaningless.)

3. Certain formulas have to be singled out as axioms.

4. Certain formal rules of inference have to be given in a finite number with the help of which new formulas can be derived from the axioms.

I add two remarks on this, namely 1. one must guard oneself against mixing ... 2. The set of formulas that in all practical cases at hand are taken as axioms is finite. This, however, is not at all necessarily required. One can consider also formal systems with infinitely many axioms. These must, then, be given through a characteristic property, so in a form: all formulas of this and this kind have to hold as axioms.

There are, now, two kinds of mathematical properties, a distinction that

[2] Example rule of substitution

plays an important role in the investigations to follow, namely what are known as properties definite with respect to decision or decidable properties and those that are not decidable. A property is called definite with respect to decision if its definition is of a kind that gives the means to decide for any object presented whether the property belongs to it or not. For example, the property to be a prime number is definite with respect to decision (for every ...). If one instead defines a property E through the stipulation: E shall belong to a number n when and only when Fermat's great theorem is correct for the exponent n, then the property is not definite with respect to decision, because the definition gives no means to hand to decide if a number has this property or not. The kind of distinction as for properties can obviously be made also for relations. With functions, the distinction corresponds to computable and non-computable functions.

It is, now, a close to self-evident condition that the property by which the class of axioms is defined must be definite with respect to decision in the case that infinitely many axioms are at hand, i.e., it must be possible to determine for each formula in a finite number of steps whether it is an axiom or not. For rules of inference, it is required in the same way that they are definite with respect to decision, i.e., to determine for each finite set of formulas whether or not the rules of inference are applicable. This holds especially for finitely many axioms and the two rules of inference.

Let now a formal system of the kind characterised be presented. We say a proposition [added above: a formula of the system] is provable in this system if it can be derived from the axioms in a finite number of applications of the rules of inference of the system. A proposition is said to be refutable when its negation is provable, and we say a formal system is complete when the propositions expressed in its symbols are either provable or refutable, or more shortly, when each such proposition is decidable.

One can, in fact, put up formal systems for certain limited partial domains of mathematical disciplines that are complete in this sense. There is, as an example, such a system for the intersection point properties of straight lines, i.e., a system in which in fact only relations of intersection of straight lines

869

are expressible by formulas and each such formula is either provable or refutable. It has been presumed from different sides that one should be able to put up such complete systems also for comprehensive mathematical disciplines, as number theory, analysis, set theory. It has been especially presumed that the formal systems already put up today for these disciplines possess this property of completeness, or at least they will have it after the addition of just a few axioms. What I would like to show now is the surprising fact that this doesn't apply and even more, namely, that it is in a certain precise sense even impossible to find complete formalizations of comprehensive mathematical disciplines, i.e., more precisely, one can show the following: If a formal system, with finitely many axioms and the rules of inference of the calculus of logic (these are substitution and the rule of implication) is at least as comprehensive as to contain the theory of natural numbers (rational number theory), it is certain that there are undecidable propositions in the system, and even undecidable propositions of the theory of natural numbers. The same holds, though, also for systems with infinitely many axioms and arbitrary other rules of inference, under a certain condition which is so general that it suffices in all cases that come into consideration in practice. Indeed, it is utterly dubitable whether it is at all possible to construct a formal system that would not satisfy this condition to which I'll return later – always under the assumption of no contradiction in the system.

This result has also a fundamental significance in so far as it shows mathematics not to be trivializable, in the sense that one could, say, find a machine that solves each mathematical problem. There cannot exist such a machine even for rational number theory.

There occurs in the theorem I have just stated the term number-theoretic proposition (I shall say briefly arithmetic proposition) that I shall now make more precise. By an arithmetic proposition shall be understood one that can be represented in the formal system of number theory put up by Hilbert and his students, I shall call it briefly $\mathfrak{Z}$. One can say briefly that they are those sentences that can be expressed with just the means of the concepts of addition and multiplication and the logical concepts.

The basic signs of the system are as follows:

$$1 \ + \ . \ = x, y, z \ldots \text{ natural numbers, } \ (x), (Ex) \ \& \lor \ \sim \ (\supset) \ \varepsilon_x \ (\,)$$

The next thing is to indicate, by choices in the above scheme, how formulas of the system are built up from these basic signs. I clarify this by some examples. One can, to begin with, distinguish between two kinds of expressions, namely those that signify numbers and those that signify propositions, for example:

$$1+1, \varepsilon_x[(Ey)\, x = y+1]$$

There occur variables in the expressions written down, but with the particularity that there occurs with each variable a variable binding operation $(x), (Ex), \varepsilon_x$. Such variables are called bound.

A variable that is not bound, so to which no sign x Ex relates, is called free. Here are, for example, written down two expressions with free variables:

$$x = 1+y, \sim(Eu, v)[x = u.v \,\&\, u \neq 1 \,\&\, v \neq 1]$$

There are also with these again two kinds. Namely, if one substitutes specific numbers for free variables, it turns into an expression without free variables, and this signifies either a number or a proposition.

870

In the first case, one calls the expression a numerical function, in the second case a propositional function or a sentence function, a one-place, two-place etc propositional function by how many variables there occur.

A numerical function associates to each number another number (or to each two numbers). A sentential function, in turn, determines a certain class of numbers, namely those numbers for which the sentence is true (or a class of pairs of numbers etc). Two complicated functions are written down here, a numerical function and a propositional function that defines the class of prime numbers. I have presented the matter here in an intuitive way. One can easily state the characteristic features of a numerical function, a sentential function, a sentence, etc.

We have now established what the formulas of the system are. Now we would have to establish what the axioms and rules of inference are. I don't want to stop further here with the axioms, they are the simplest properties of + and $\cdot$ and a number of purely logical propositions. Two rules of inference are used, namely those two which one has actually always found to suffice, namely the rule of implication and the rule of substitution:

The rule of implication states that whenever a formula $\mathfrak{A}$ and whenever further a formula of the shape $\mathfrak{A} \supset \mathfrak{B}$ have been proved, the formula $\mathfrak{B}$ can be inferred from these.

The rule of substitution states that whenever a formula of the shape $(x)F(x)$ has been proved, i.e., a formula that begins with the sign ( ), one can then leave out the all-sign and substitute for x an arbitrary numerical expression (an expression that signifies a number).

One can easily express in the system $\mathfrak{Z}$ all the concepts and theorems of rational number theory, and even of algebraic number theory, as long as they are not abstract, and conduct the proofs of these theorems from the axioms.

We shall next say of an arbitrary class $\mathcal{K}$ of entire numbers that it is *expressible in the system* $\mathfrak{Z}$, *or contained in the system* $\mathfrak{Z}$, if there is a propositional function $F(x)$ with one free variable such that $x \,\varepsilon\, \mathcal{K} \equiv F(x)$, analogously for relations, that is to say for classes of pairs of numbers. We say similarly of a numerical function of natural numbers $f(x)$ that it is contained in the system $\mathfrak{Z}$ if there is in $\mathfrak{Z}$ a numerical function with one free variable that takes on always the same value as $f(x)$. So one can say in this sense that the class of prime numbers or the function smallest common multiple are contained in $\mathfrak{Z}$.

It is certain that not all classes and relations are contained in $\mathfrak{Z}$, because there are clearly more than denumerably many relations but just denumerably many formulas in $\mathfrak{Z}$. One can easily give examples of functions that are not contained in $\mathfrak{Z}$, namely through processes that are analogous to the diagonal process. Nevertheless, *and this is important for what follows*, one cannot give any *decidable properties* and also no *computable functions* that would not be contained in $\mathfrak{Z}$. One cannot prove, say, that

871

II Decidable

each property or function [cancelled: of the said kind] must be contained in $\mathfrak{Z}$, but one can show that all procedures for the definition of decidable properties and computable functions (especially the procedure of recursive definition) lead always only to functions that are contained in $\mathfrak{Z}$. It is presumed from different sides that one would never be able to find any kind of new procedures that would lead to decidable properties not contained in $\mathfrak{Z}$. Even more holds for decidable properties (as said, at least for those known today), namely the following:

Let $\mathcal{K}(x)$ be such a decidable property. There exists then in $\mathfrak{Z}$ an equivalent propositional function $F(x)$. But it holds even that if $\mathcal{K}(a)$ holds for a specific natural number a, then $F(a)$ is provable in $\mathfrak{Z}$. That is, ... Example prime number.

To say that a formal system S contains number theory shall, then, mean that each sentence, each propositional function and numerical function that is expressible in $\mathfrak{Z}$ is also expressible in S, and that each proof that is conductible in $\mathfrak{Z}$ is also conductible in S. So this is what we must assume about the system S in what follows, to be able to prove our theorem.

I define next a few concepts that pertain to an arbitrary formal system S. There are in S certain *basic signs*. Certain combinations of these basic signs are the *formulas*. A subclass of formulas form the *propositions* (not every formula is a proposition). A subclass of the propositions is, next, the *axioms*. Further, a certain relation between formulas is given through rules of inference. This relation shall be called *immediate consequence*, i.e., more precisely, a formula $\mathfrak{A}$ shall be called an immediate consequence of some other formulas $\mathfrak{B}, \mathfrak{C}$ (finitely many) if $\mathfrak{A}$ results through a *single application* of a rule of inference of the system to the formulas $\mathfrak{B}, \mathfrak{C}$, etc. For example, if three formulas of the shape $\mathfrak{A}$, $\mathfrak{B}$, $\mathfrak{A} \rightarrow \mathfrak{B}$ are at hand, then the second is an immediate consequence of the first and last, for it can indeed be recovered through the application of the implication rule to the first and last.

In the case that the implication rule is the only rule of inference of a formal system, the relation of immediate consequence would obtain only between three formulas of the correct shape, i.e., the statement that $\mathfrak{B}$ is an immediate consequence of $\mathfrak{A}$ and $\mathfrak{C}$ would mean that $\mathfrak{C} = \mathfrak{A} \rightarrow \mathfrak{B}$.

We can now define what is to be understood by a *proof or a chain of inference*. A proof is a finite series R of formulas of the system S with the following property: Each formula of the series shall be either an axiom or an immediate consequence of preceding formulas of the series. A series of formulas R shall be, further, called a *proof for* a formula $\mathfrak{A}$ if:

1. the series R is a proof and
2. the last formula of R is $\mathfrak{A}$

The concept *proof for* is, then,

872

a relation between a finite series of formulas and a formula.

I have now defined a series of concepts that pertain to the formulas of a formal system. One calls such concepts metamathematical. All of the concepts defined so far have one property in common, namely, they are definite with respect to decision. Let us take, for example, the concept *chain of inference or proof*. To decide of a series of formulas at hand whether it is a proof, one has to go through the formulas one after the other and to determine for each formula whether it is an axiom or an immediate consequence of the preceding ones. The first determination is naturally to be done in finitely many steps (as assumed), and even the second requirement has a finite number of trials as only finitely many formulas precede the one in question (example with substitution rule and implication rule). The relation "proof for" is, correspondingly, a relation definite with respect to decision.

A formula φ is called *provable* if there exists a series of formulas that is a proof for φ. This relation is, contrary to all of the previous ones, not definite with respect to decision.

There exist in each formal system only denumerably many formulas, for there exist only finitely many basic signs and each formula is a finite combination of these basic signs. Therefore the formulas can be numbered through, i.e., one can associate in a one-to-one way to each formula a natural number, say lexicographically or whatever other way. The essential thing for what follows is just that the numbering has been so arranged that one can actually determine for each formula the associated number and conversely for each number the associated formula, something that naturally is the case with lexicographical ordering. So we think some numbering for the system S at hand as chosen and kept fixed in what follows. It is in virtue of this one-to-one association between formulas and natural numbers that (something that is crucial for what follows) there is associated to the metamathematical concepts at hand (classes and relations of formulas) a certain class or relation between natural numbers.[3] For example, to the class of sentences of the formal system considered is associated the class of those natural numbers that are *numbers of sentences*, or to the relation of immediate consequence is associated the relation R between natural numbers that is defined as follows:

R shall obtain between the numbers $n_1\, n_2 \ldots n_k$ if and only if the relation

[3] [The sentence has many changes, but the original is clearly readable: It is in virtue of this numbering that one can now (something that is crucial for what follows) associate to each of the previously defined metamathematical concepts (i.e., classes and relations) a certain class or relation between natural numbers.]

of immediate consequence obtains between the numbers $n_1 n_2 \ldots n_k$, i.e.,

if the last of these formulas is an immediate consequence of the rest.

There are, further, only denumerably many formulas at hand, and therefore there exist also just denumerably many proofs, for each proof is a finite series of formulas. Therefore one can enumerate the proofs as well, and thereby there arises the relation:

The series of formulas a is a proof of the formula b corresponds to the following relation between two *natural numbers*: —

We obtain therefore in this way a number of relations and classes of natural numbers, and it is clear now that these relations and classes are just as definite with respect to decision as the corresponding metamathematical concepts from which they have arisen. I.e., if two determinate numbers a, b are laid at hand, one can

873

always determine in a finite number of steps whether a is the number of a proof for the formula with the number b. For one can produce, to start with, the proof with the number a and then the formula with the number b, and then it needs only to be determined whether the formula in question is the last formula of the proof in question.

The classes and relations of natural numbers we have thus defined are, admittedly, obtained by a detour through metamathematical concepts, but they don't differ in the least from the classes and relations between natural numbers that are otherwise encountered in mathematics. I.e., more precisely, all of these number concepts are contained in the system $\mathfrak{Z}$ (i.e., expressible through the basic symbols of the system $\mathfrak{Z}$), i.e., more precisely, if the system S contains only finitely many axioms and if its rules of inference are the usual ones (i.e., rules of substitution and implication), then the concept xBy and even the further ones are surely contained in $\mathfrak{Z}$. In case the system contains infinitely many axioms or other rules of inference, then xBy and the other corresponding concepts are surely contained in $\mathfrak{Z}$, if the number concepts that correspond to the class of the axioms and to the relation of immediate consequence are contained in $\mathfrak{Z}$. That amounts, by what was said earlier, in all known cases to the same as when one says that the class of axioms and rules of inference are to be definite with respect to decision. If one, now, requires of a formal system S that 1. both of the requirements presented are satisfied (these were either finitely many axioms and usual rules of inference, or axioms and rules of inference [written above: definite with respect to decision] expressible in $\mathfrak{Z}$) and that 2. S contains the system $\mathfrak{Z}$, then there results the strange fact that the metamathematical concepts that pertain to formulas of S (and their arithmetical correlates) are once again expressible through formulas of S. For they are expressible by formulas of $\mathfrak{Z}$ to start with, and $\mathfrak{Z}$ has to be contained in S. This is, as you shall see, the decisive condition for the proof that follows. All conditions for the proof to follow are especially fulfilled by the system $\mathfrak{Z}$, further for all formal systems of analysis and set theory put up so far.

I shall now go over to the proof. Let, then, S be a system that satisfies the conditions and contains the system $\mathfrak{Z}$. I consider the two sequences of concepts:[4]

$x\mathfrak{B}y$ $\varphi(xy)$ | definite with respect to decision – computable

numerical relation numerical function

contained in $\mathfrak{Z}$ therefore in S

$x\mathfrak{B}y$ propositional function in S

$F(xy)$ numerical function in S

874 [This is the page with formulas and a few words in longhand.]

$xBF(yz)$ in S definite with respect to decision therefore

(1) $aBF(bc) \supset Bew\, aBF(bc)$ negation analogously

Let the following be formed:

(2) $\overline{(Ey)}yBF(xx)$ number p (because in S) [above: is a specific number]

(3) $\overline{(Ey)}yBF(pp)$ number $F(pp) = q$

(3) claims unprovability of the formula with number $F(pp)$ which it itself [is]

(3) is the sentence (3) is undecidable

A. Assume (3) is provable proof № k

[4] [The text from now on is written in longhand.]

then we have $kBF(pp)$

therefore because of (1) even provable

therefore also $(Ey)yBF(pp)$ [second y added] provable *contradiction*

So (3) not provable if W [widerspruchsfrei]

$W \rightarrow \overline{(Ey)}yBF(pp)$

(3) correct if W

B. (3) correct because free from contradiction, i.e., for each a:

$\sim aBF(pp)$ therefore provable for each a

If now, on the other side, $(Ey)yBF(pp)$ provable then inconsistency

[The following is found on the page before the main text.]

$x \varepsilon K \equiv F(x)$

$K(a) \supset Bew[F(a)]$ [right bracket added]

Propositional axiom, immediate consequence, proof, proof of provability

$x\mathfrak{B}y$ proof № x is a proof for the formula with number y

$f(xy)$ = number of the formula that arises from formula № x when the number y is substituted in place of the free variables

$xBF(yz)$

(1) $aBF(bc) \supset Bew[aBF(bc)]$

$\sim[aBF(bc)] \supset Bew[\sim(aBF(bc)]$

(2) $\sim(Ey)yBF(xx)$ number p

(3) $\boxed{(Ey)yBF(pp) \text{ number } F(pp) \quad F(pp) = q}$

(4) $kBF(pp)$ provable

(5) $(Ey)yBF(pp)$

(6) $W \rightarrow (\overline{Ey})yBF(pp)$

(7) $\sim aBF(pp)$

875

III

A further strange result follows from what has been presented so far, concerning proofs of freedom from contradiction. The statement that a formal system *S* is free from contradiction is a metamathematical one. It claims that of two formulas $\mathfrak{A}$ and $\sim\mathfrak{A}$, both can never be provable. The metamathematical concepts can be replaced by arithmetic ones by the procedure presented earlier, and therefore one can express the statement of freedom from contradiction through a certain arithmetic proposition. More precisely, one can give an arithmetic proposition that is equivalent to the freedom from contradiction of the system in question. This proposition is, as an arithmetic one, contained within the system *S* itself, the freedom from contradiction of which it expresses. From the above proof follows at the same the result that this statement of freedom from contradiction can never be provable within the system *S*, the freedom from contradiction of which it claims. That is to say, in other words, one can never prove the freedom from contradiction of a formal system (that satisfies the above conditions) with the help of ways of inference formalizable in this system, but one needs to use for that always some ways of inference that go beyond the system, not contained in it. This fact results at once formally from the above. For we have shown that if the system is free from contradiction, then the undecidable proposition is correct. A proof of this fact can, as a more detailed investigation shows, be carried formally through from the axioms of the system $\mathfrak{Z}$, therefore also in *S*, i.e., one can prove this formally in *S*. If one could, now, prove the proposition *W* in *S*, one could also prove the proposition , whereas it was just shown that this proposition is undecidable. As is known, Hilbert and his students are looking since long for a proof of freedom from contradiction for the formal system of classical analysis and set theory, and they search this proof from as weak assumptions as possible. As far as possible, only the simplest [added above: combinatorial] facts [cancelled: of spatial intuition] must be used. It follows by the result described that all these attempts to conduct a proof of freedom from contradiction with so weak means of proof are condemned to fail, because all purely combinatorial facts are clearly expressible in the axiom system of analysis. To carry through a proof of freedom from contradiction, one has to absolutely use proof methods that go much further than has happened with the Hilbertian ansatz so far.

To end with, I would like to point out that with all the things I have presented so far, the question has always been about *relative* unprovability and relative undecidability, i.e., of unprovability and undecidability in a specific formal system. One can extend each such system by the addition of a few sentences, namely sentences that are pertinent in such a way that previously undecidable sentences are decidable in the new system. It is, for example, sufficient to take as a new added axiom the statement of the freedom from contradiction of the old formal system (i.e., more precisely, the arithmetic proposition equivalent to it). There exist in the new systems naturally again undecidable arithmetic propositions that require a new extension and so on into the transfinite. One has to say into the transfinite because it would not be sufficient, say, to take denumerably many extensions, for even after denumerably many extensions, there would remain undecidable propositions for an $\omega + 1$ extension.

876 [The upper left corner has four seemingly unrelated words.]

So this is to say, one can in fact give for each undecidable sentence constructed in this way a further axiom system in which it is decidable, but one can give no system in which each sentence is decidable.

One can carry out yet *another kind of extension* that makes the previously undecidable propositions decidable as well, one that has great mathematical interest, namely through the introduction of higher types. I illustrate this in axiom system 3. If one adds here as new concepts those that stand here, ones that correspond to the axioms for classes and relations of numbers, the propositions that were previously undecidable become decidable. The transition into the concepts of classes and relations between numbers means, however, a transition from number theory to analysis, for real numbers are defined as sequences of rational numbers or as classes of rational numbers. The strange state of things follows that there exist number-theoretic problems that are not solvable in an axiom system of number theory, but only in one of analysis, which then shows that analytical number theory can be indispensable in certain cases. [Cancelled: Perhaps I return to these questions again later.]

On the impossibility of proofs of freedom from contradiction

Mathematical logic 22nd seminar hour 4.7.1932

Gödel:

The following enter into Herbrand's proof for the consistency of arithmetic:

a) basic constant o

b) a one-place operation "+1"

c) the basic relation "="

d) axioms 1) $x = x$

2) $x = y \,.\supset.\, y = x$

3) $x = y \,.\, y = z \,.\supset.\, x = z$

4) $x = y \,.\equiv.\, x+1 = y+1$

5) $\sim .\, x+1 = \mathrm{o}$

An infinite set of new axioms is introduced through complete induction, say for functions of one variable through the stipulation

$f(\mathrm{o}) = \alpha$ here a constant, i.e., a number $\mathrm{o}, 1, 1+1, 1+1+1, \ldots$

$f(x+1) = \beta(f(x))$ here a function of one variable already known.

If one puts in evidence further variables,

$f(\mathrm{o}; x_1, x_2, \ldots x_n) = g(x_1, x_2, \ldots x_n)$

$f(x+1; x_1, \ldots x_n) = h(f(x; x_1 \ldots x_n), x_1, x_2, \ldots x_n)$

As an example $f(x,y)$,

$f(x,0) = x$

$f(x,y+1) = f(x,y) + 1$

Quite generally, one can add (meta-arithmetically) any arbitrary system of equations, whenever it allows the computation in a unique manner and each natural number appears there, then even, for example, the recursion

$u(n, a, \mathrm{o}) = a$

$u(\mathrm{o}, a, b) = a + b$

$u(n+1, a, b) = u(n, a, u(n+1, a, b-1)$

This prescription does not allow, though, to determine of a formula whether it is an axiom or not.

There comes finally in addition the principle of complete induction.

(Hahn:) Not as with Peano, as an axiom, but instead separately for each function as an axiom of its own,

$$\varphi\,\text{o}:.\,(x)\,.\,\varphi\,x \supset \varphi\,x+1\,:\supset.\,(x)\varphi\,x$$

The proof of freedom from contradiction is now carried through for each finite subset of the axioms. To be able to carry it through, only functions φ without bound variables are allowed to enter in the inductive statement. Let then, be it $\varphi\,a$, be it $\sim\varphi\,a$ derivable from the axioms. That is to say, the recursively defined formulas $\varphi_1 \ldots \varphi_n$ are present, and then a) to d) to be shown, then the logical product

$$H\,.\,(u)A_1(u) = B_1(u)\ \ldots (u,v)A_1'(u,v) = B_1'(u,v)\ \ldots$$
$$(z)(\exists x) :.\, \varphi_1(\text{o})\varphi_1(x) \supset \varphi_1(x+1) \supset \varphi_1(z) \ldots$$

Here H is the logical product of the axioms *sub* d), and a finite number of formulas for one, for two, for more than two variables occur, finally the formulas φ_1 to φ_n for which complete induction is stated, brought here already to normal form with the help of the rules of passage.

One can now specify the construction of the infinite field, i.e., to each natural number n a field of order n in such a away that the substitution of the field functions in the functions of the formulas give the truth value true. To this end, the reduction is built that consists here simply in introducing in the last part the corresponding index functions, i.e., $(z)(\exists x)$ delivers $f_x(z)$.

Mathematical logic, 22nd seminar hour, page 2 4.7.1932

The reduction obtains hereby the shape

$$H\,.\,A_1(u) = B_1(u)\ \ldots A_1'(u,v) = B_1'(u,v)\ \ldots$$
$$\varphi_1(\text{o})\varphi_1(f_x^1(z))\,.\supset.\,\varphi_1(f_x^1(z)+1)\,.\supset.\,\varphi_1(z)\ldots$$

Here there occur just $+1, \varphi_1 \ldots \varphi_k\ f_x^i$ o =

One can specify with the natural numbers, o included, so o, 1, 2..., an infinite field and with it a sequence of finite fields the elements of which, substituted by corresponding rules of associations, make the expression always true.

Then, for example,

0	o
$+1$	the successive number
φ_i	number that results after computation of the recursion formula
$f_x^i(k)$	Dilemma

I $\varphi_i(\text{o})$ false or $\varphi_i(k)$ true, then $f_x^i(k) = \text{o}$

II $\varphi_i(\text{o})$ true and $\varphi_i(k)$ false, then $f_x^i(k) = a$

This a gets determined so that there exists in the sequence of natural numbers an a such that

$$\varphi_i(a) \text{ is true and } \varphi_i(a+1) \text{ is false}$$

$a = b$ shall be true if arithmetically $a = b$
false if arithmetically $a \neq b$

In virtue of the association found, it is necessary to furnish a proof only for the inductive axioms. One uses for this in the case,

$$\text{I} \quad \frac{\dfrac{\dfrac{\underset{f}{\varphi_1(\text{o})} \;.\; \overline{\varphi_1(\text{o}) . \supset . \varphi_1(1)}^{\,}_{w}}{f}}{w} \quad w}{w} \quad :\supset \varphi_1(k)$$

[Added at right: There enters here essentially the restriction for the φ and the recursive definition.]

and

$$\text{II} \quad \frac{\dfrac{\underset{w}{\varphi_1(\text{o})} .\; \dfrac{\underset{w}{\varphi_1(a)} . \supset . \underset{f}{\varphi_1(a+1)}}{\underset{w}{} \quad f}}{f} \quad \supset .\underset{f,\ f,\ f}{\varphi_1(k)}}{w}$$

(Menger:) What is the sense of a theory in which all natural numbers enter through complete induction?

(Gödel:) All natural numbers enter, sure, in the theory, but not instead the *tertium non datur,* $(x)\varphi(x).\vee.(\exists y)\sim\varphi(y)$ that is a provable formula. I.e., one can derive consequences from the proposition: "The Fermat conjecture is correct, or it is not provable." One cannot, though, conclude its correctness in this theory on the basis of complete induction, i.e., "correct for $z = 3$" and "if correct for k, then correct for $k+1$,"because of the restriction that has been made.

Gödel:
On the impossibility of proofs of freedom from contradiction.

A formal theory will be, as is well known, described as complete when for each meaningful proposition P either P or its negation $\sim P$ is derivable from the axioms

But now, each formal system in which there occurs addition and multiplication contains propositions that are undecidable.

Mathematical logic, 22nd seminar hour, page 3 4.7.1932

Let such a formal system S contain, then,

what are known as variables of type 1, $x, y, z \ldots$

and it is required

IA. that $x+y, x.y$ are variables of the same type, that especially 1 is of the first type.

IB. the propositions by which = is an *aequitas* (reflexive, symmetric, transitive)

for "+" (without 0!) $x+1 \neq 1 \quad x+1 = y+1 \,.\rightarrow.\, x = y$ and $x+(y+1) =$
$(x+y)+1$ and analogously for "."

IC. the induction

from $\varphi(1)$ and $\varphi(x) \rightarrow \varphi(x+1)$, $(x)\varphi(x)$ can be inferred

I.e., even arithmetic is formally contained in the theory.

II. A sharper formulation of freedom from contradiction.

$1, 1+1, 1+1+1, \ldots$, briefly denoted by a, are CIPHERS.

If for $\varphi(x)$: $\varphi(1)$ $\varphi(1+1)$ ... $\varphi(a)$... are provable

then $(\exists x) \sim \varphi(x)$ is not provable.

I.e., it will be certainly said that something is wrong then, but not that a formal contradiction is derivable. In other words, even if the formula is provable for each single n, it does not mean the same as the provability of $(x)\varphi(x)$.

III. (can be extended later:) there are infinitely many axioms.

We say now of a class of natural numbers K that it is contained in the formal system S if there exists a $\varphi(x)$ – where x is of type 1 – so that

if $a \, \varepsilon \, K$, then $\varphi(a)$ provable

if $\sim a \, \varepsilon \, K$, then $\sim \varphi(a)$ provable

Analogously for relations as classes of ordered pairs of natural numbers

if aRb, then $\psi(a, b)$ provable

if $\sim aRb$, then $\sim \psi(a, b)$ provable

As an example, say, the class of even numbers [G = Gerade]

(Formula:) $(\exists y)\, x = y + y$
free variable $\quad \widehat{=}$

$$
\begin{array}{lll}
2 \, \varepsilon \, G & (\exists y) & 1+1 = y+y \\
\sim 3 \, \varepsilon \, G & \sim (\exists y)\, 1+1+1 = y+y &
\end{array}
$$

or the relation of x smaller than y

(Formula) $(\exists z)\, y = x + z$ [the manuscript has $x = y + z$]
free variable $\quad \widehat{=} \quad \widehat{=}$

$$
\begin{array}{ll}
1 < 2 & (\exists z)\, 1+1 = 1+z \\
\sim (2 < 1) & \sim (\exists z)\, 1 = 1+z
\end{array}
$$

or "prime number"

$$
\sim (\exists uv)\, u < \underset{\widehat{=}}{x} \,.\, v < \underset{\widehat{=}}{x} \,.\, \underset{\widehat{=}}{x} = uv
$$

There are only denumerably many formulas, but definitely nondenumerably many classes of natural numbers that therefore cannot be contained.

It can be shown that each recursively defined sequence is contained, possibly even each decidable formula.

The set of basic signs is denumerable, and therefore also the set of meaningful expressions composed from them, i.e., a one-to-one association with the natural numbers is possible,

Mathematical logic, 22nd seminar hour, page 4 4.7.1932

for example $\sim$ v $(\exists x)$ etc.
1 2 3

Each expression is now composed of a finite sequence of such signs, i.e., fixes a series of associated numbers $e_1, e_2 \ldots e_n$. If one takes these next as exponents of the prime numbers in an ascending order, each expression itself obtains again an associated natural number

$p_1^{e_1} p_2^{e_2} \cdots p_n^{e_n}$

By this, not to each number is associated a formula, but to each formula is associated – because of the prime decomposition in a unique way – a number.

In this class of numbers M, we have for example:

Formulas that begin with the sign of negation distinguished as numbers that are divisible precisely by 2, $2^1 p_2^{e_2} p_3^{e_3} \ldots$

The relation "*a* is longer than *b*" between formulas, i.e., consists of more signs, [cancelled: through $xRy\ u, zw$, this is precisely the number of base numbers]

If ordered series of formulas are again given in the same way,

$[F_1 F_2 \ldots F_n]$

a number $p_1^{e_1} p_2^{e_2} \ldots$ can be uniquely associated to each series of formulas, because the formula numbers are uniquely determined by the same procedure.

Next a series of formulas is designated as a "proof" if the formulas that occur there first are axioms and if each formula that follows is derived from the previous ones though the rules of inference.

By the above, there corresponds to the subclass of proofs among the class of series of formulas a subclass from the class of numbers of series of formulas.

We say: F is a proof for formula g if

a) F is a proof $\qquad$ x the number of proof F

b) the last formula of F is precisely g $\qquad$ y the number of formula g

Hereby a relation $x\mathfrak{B}y$ for natural numbers is now fixed: "x is the number of a proof for the formula with the number y," a relation that is decidable in a finite number of steps and moreover as a recursive one contained in the formal system.

That is, then, there is in the formal system S an expression $\psi(x,y)$ such that

$a\mathfrak{B}b$ $\qquad$ then $\psi(a,b)$ provable (in short: *bew*) $\quad \psi(a,b)$

$\sim(a\mathfrak{B}b)$ $\qquad$ then $\sim\psi(a,b)$ provable

For the axioms, it is of course required that it is decidable (in finitely many steps) whether a formula at hand is an axiom, i.e., that axioms just be recursively defined, or what amounts to the same, $\mathfrak{B}$ must be recursive.

Next, the operation $\Pi(g,z)$ associates to the formula g the one that one obtains when the fixed number z is substituted for the free variable of type 1. There corresponds to this operation a relation $\varphi(x,y)$ between natural numbers, the number of the formula that one obtains from formula № x through substitution of the cipher y in place of the free variable.

Mathematical logic, 22nd seminar, page 5 $\qquad$ 4.7.1932

All the auxiliary means for the construction of an undecidable proposition have now been put together.

The formula $x\mathfrak{B}\mathfrak{F}(yy)$ belongs, because of its recursive definition, to the formal system, i.e., if

(3) $\quad a\mathfrak{B}\mathfrak{F}(bb)$ $\qquad$ then $\quad \varphi(a,b)$ provable

(4) $\sim a\mathfrak{B}\mathfrak{F}(bb)$ $\qquad$ then $\sim\varphi(a,b)$ provable

We form now:

(1) $\sim(\exists y)\varphi(y,x)$ $\qquad$ and put therein for the free variable x the number p the formula itself has:

(2) $\sim(\exists y)\varphi(y,p)$ $\qquad$ the formula thus arrived at has the number $\sigma(p,p)$

(2) is now a proposition without free variables. We infer now indirectly:

Were next

I $\sim(\exists y)\varphi(y,p)$ provable, the number of its proof k, but then $k\mathfrak{B}\mathfrak{F}(p,p)$, so by the previous (3), $\varphi(k,p)$ would be provable, and the rule of generalization 2 applied on it, $(\exists y)\varphi(y,p)$ [manuscript has $\varphi(x,p)$]. I.e., the negation of (2) would follow, the system would be contradictory, whereas it is required to be free from contradiction. The proposition itself is, then, not provable.

$W. \supset . \sim bew[\sim (\exists y)\varphi(y,p)]$

Were next

II $(\exists y)\varphi(y,p)$ provable, i.e., also the negation of (2) and let us have for a specific k, say

$\sim k\mathfrak{B}\mathfrak{F}(p,p)$

Then $\sim \varphi(k,p)$ is provable by (4).

If on the other hand $k\mathfrak{B}\mathfrak{F}(p,p)$ holds for a specific choice, it would mean that the formula with number $\mathfrak{F}(p,p)$ is provable, i.e., precisely $\sim (\exists y)\varphi(y,p)$ in contradiction with assumption II, so hereby even $\sim \varphi(k,p)$ would be provable. I.e., then the sharper formulation of freedom from contradiction, as required for the system, would not hold, because by the above, from $(\exists y)\varphi(y,p)$ would follow $\sim \varphi(k,p)$ for each k.

Each contradiction can be derived or reduced back to the contradiction $1 \neq 1$ and therefore, if this formula carries the number q, one can give to the conclusion under I the specific form:

$\sim(\exists x)x\mathfrak{B}q . \supset . \sim(\exists x)x\mathfrak{B}\mathfrak{F}(p,p)$

I.e., if there exists no proof of whatever number x for the formula Nº q, then there exists as well no x that would be the number of a proof for $\mathfrak{F}(p,p)$.

$\mathfrak{B}$ and $\mathfrak{F}$ are two-place relations between natural numbers, and the whole an arithmetic proposition to which there is associated in the formal system the proposition

$\sim(\exists x)\psi(x,q) . \supset . \sim(\exists x)\varphi(x,p)$

But this is the arithmetic expression for the statement that "the formal system is free from contradiction," i.e., the statement as a formula of the system S, "formalized in S." Were the formula, now, provable, it would follow that the system is contradictory. The proposition that a formal system is free from contradiction cannot be proved within the same system, for otherwise the system would be contradictory. But this means that the concept of decidability is a relative one. One could prove trivially an undecidable proposition in an extended theory by adding it, say, as an axiom, but the system achieved in this way would not possess the property of freedom from contradiction in a sharpened sense anymore.

The existence of undecidable propositions in any formal system containing arithmetic

0

I appreciate very much my having the opportunity of speaking before you on some of the modern results in the foundations of mathematics and I hope I shall succeed in making leading ideas clear to you, despite the very technical character of the work.

The subject I want to talk about is closely connected with the so-called formalisation of mathematics, i.e., with the fact that all mathematics and logic (at least all mathematics and logic that has been developed so far) can be deduced by means of a few axioms and rules.

1.

The modern investigations in the foundations of mathematics gave as one of their outstanding results the fact, that all mathematics and logic (at least all mathematics that has been developed so far) can be deduced by means of a few axioms and rules of inference.

In order to bring out this fact clearly it was necessary at first to replace the imprecise and often ambiguous colloquial language (in which mathematical statements are usually expressed) by a perfectly precise artificial language the logistic formalism. This formalism consist of a few primitive symbols which represent the primitive notions of logic and mathematics and play the same role as the words in ordinary language. I wrote some examples of the primitive symbols on the blackboard. ——————— Now any logical or mathematical proposition can be expressed by a formula composed of these primitive symbols and vice versa any formula composed of our

2.

primitive terms according to certain rules (which constitute the grammar of our logical language) expresses a definite mathematical statement. In practice it would be very inconvenient to express mathematical statements in this way by means of the primitive terms of ... i.e. our formulas would become very long and cumbersome therefore besides our primitive terms new symbols are introduced by definitions but it is to be noted that this (device) serves merely the practical purpose of abbreviation and therefore is entirely dispensable from the theoretical point of view since we can replace in every formula the new symbols by their meaning expressed in the primitive terms. So we may disregard the possibility of introducing new symbols by definition and think of any mathematical statement as expressed by our primitive terms alone. —— The process of deduction, i.e., of proof is represented in our formalism in the

3.

following manner: Some of our formulas are considered as axioms, i.e., as the starting point for developing mathematics and in addition to that certain rules of inference are stated which allow one to pass from the axioms to new formulas and thus deduce more propositions. One of the rules of inference, e.g., reads. If A and B are two arbitrary formulas and if you have proved the formula A and $A \rightarrow B$ you are entitled to conclude B. The other rules of inference are of a similar simple character. In practice all of them are purely formal, i.e., they do not refer to the meaning of the formulas but only to their outward structure and so they could be applied by someone who knows nothing about the meaning of the symbols. One could even easily device a machine which would give you as many correct consequences of the axioms as you like, the only trouble

4.

would be that it – at random and therefore not the results one is interested in. By iterated application of the rules of inference starting from the axioms we obtain what I call a chain of inference. A chain of inference is simply a finite sequence of formulas $A_1 \ldots A_n$ which begins with some of our axioms and has the property that each of it's other formula's can be obtained from some of the preceding ones by applications of one of our rules of inference. Instead of chain of inference I shall also use the term formal proof or briefly proof. A proof ending up with the formula F is called a proof for the formula F and of course we shall call a formula F provable if there is a proof for it, which means the same thing as: F can be obtained from the axioms by iterated application of the rules of inference. A symbolism for which

5.

axioms and rules of inference are specified in the manner I have just described is called a formal system and the fact to which I referred in the beginning of my talk can now be expressed by saying that one has succeeded in reducing all of mathematics and logic to a formal system [added above: in such a way that every mathematical proof can be]. Owing to this fact certain general questions concerning the structure of mathematics which formerly had to be left to vague specifications (and could not even be stated precisely), have become amenable to scientific treatment. I want to deal with two of these questions. The first concerns the freedom from contradiction of mathematics. This question can now be stated in a perfectly precise way as follows. "Does there exist any formula A such that A and not $\sim(A)$ are both provable" where the term provable has the precise meaning which I defined before by our rules of inference in a finite number of steps.

6.

It can be easily shown that if there existed two formulas A and $\sim A$, both of which were provable, then any formula whatsoever would be provable for instance also the formula $0 = 1$. So it is of vital importance for our formalism that his should not happen and the problem of giving a proof that it cannot happen arises. But at the same time an objection can be brought against the soundness of this problem. Namely one may say: Suppose we had given a proof for consistency then owing to the fact that it is a mathematical proof it must necessarily proceed according to the axioms and rules of inference for mathematics and logic. So in order to be convinced by this supposed proof we must know that our axioms and rules of inference which we used always lead to correct results. But if we know this in advance then no proof for freedom from contradiction is necessary

7.

(because rules of inference which lead to correct results cannot lead to A and $\sim A$ because these two formulas cannot both be correct). Fortunately the actual situation is slightly different. For mathematics consists of two distinct parts which are usually referred to as finite and transfinite mathematics and which may be roughly characterised as follows. Under the first heading (of finite mathematics) are comprised all such methods of proof which do not presuppose the existence of any infinite set whereas under the second heading (of infinitary mathematics) fall those methods of proof which do presuppose the existence of infinite sets and are based on this assumption. (e.g., let P be any arithmetical proposition and let's consider the statement either every integer has the property P or there is an integer which has not this property.

8.

Now nobody has ever questioned seriously the consistency of finite mathematics whereas the situation is quite different with the transfinite methods based on the assumption of the existence of infinite sets, which by the way is by far the greater part of mathematics now existing. In this domain of mathematics actual contradictions had arisen unexpectedly by toward the end of the 19. century the so called paradoxes of the theory of aggregates. In order to avoid them certain restrictions on the previous assumptions concerning the existence of infinite sets had to be made. These restrictions can be made in a very natural way and they do not affect in any way the mathematical results previously obtained, but nevertheless the faith of many mathematicians in the transfinite methods

9.

was shaken by this bad experience and there remains the fear that other paradoxes may arise in spite of the restrictions. Now I think it is clear what the question of proving freedom from contradiction really is about. It is the problem of proving the freedom from contradiction of transfinite mathematics by means of finite methods, i.e., using in the proof for consistency only such methods as are not based on the existence of infinite sets. So much for the meaning of the 1. problem, the question of consistency. The second problem is in its treatment so closely related to the first that it can hardly be dealt with separately. It is the question of completeness of the formal system for mathematics, i.e., the question whether every mathematical statement expressed by a formula of the system can be decided (either in the affirmative or in the negative) by means of the rules of inference and axioms, i.e., is it

10.

true that if A is any arbitrary formula expressing a proposition then either A or $\sim A$ is provable? Or are there formulas for which neither one of the two is provable? I am going to sketch a proof which answers both questions in the negative in the following sense: 1.) It is not possible to prove one system consistent, using only a part of the methods of proof embodied in its axioms and rules of inference. In fact, it is not even possible to prove it consistent using all of its methods of proof. 2.) There are propositions in fact even propositions belonging to the arithmetic which cannot be decided by a formal proof. Of course mathematics can be formalized in different ways, i.e., the axioms and rules of inference representing mathematics can be chosen in different manners and so one may suspect that our two results depend on the special system for mathematics which we chose. But this is not the case. It can be shown that

11.

the two theorems which I just stated hold good whatever formal system we may choose provided only that arithmetic of integers in its usual form is contained in the system and that no false arithmetic statement is provable, i.e., the axioms and rules of inference should not lead to results which can be disproved for intuitive reasons.

The proof for these two statements (impossibility of a proof for consistency and existence of undecidable propositions) is very cumbersome if worked out in all details but I hope to succeed in making the leading ideas clear to you.

12.

Suppose system given

Among expressions also such as x is > 6
not proposition but becomes so if substituted
called propositional function
expresses properties
Similarly with several properties expressing relations

Let[1] primitive symbols of the system be $\sim, \rightarrow, E, x, r_s, \ldots s_n$
Any formula = combination of primitive symbols = sequence
Therefore numbering possible

[1] Changed into: Let's write down the

in many ways, we choose the following:
(number primitive symbols ...)
Proof = sequence of formulas = sequence of numbers
Numbering of proofs
Not all numbers used up but one to one

Owing to numbering: class of formulas ⇒ class of numbers
relation of formulas ⇒ relation of numbers
e.g., relation of being longer
Similarly for any relation (called metamathematical) ⇒
relation arithmetic

13. *12a*

A relation between formulas such as being longer is
Analogous to analytical geometry (also statements)

Further examples needed for subsequent proof.

Relation of immediate consequence $\frac{P \quad Q}{R}$ means P is the formula $Q \rightarrow R$ (= implication with Q as first and R as second term)

What does that mean for corresponding numbers p, q, r

Series of exponents correspond to series of symbols therefore series of exponents of p must be composed of those for q and r with one between them

purely arithmetic relation
call it for the moment derived
for any three numbers it can be ascertained whether or not

Arithmetic definition of the integers which are numbers of proof as follows:

Suppose n axioms with numbers $k_1 \ldots k_n$ (definite numbers which can be computed)

Suppose further only one rule of inference which makes no essential difference

14.

Recall definition of formal proof (or chain of inference)

What does that mean for numbers?

According to correspondence if n is number of proof the exponents numbers of formulas occurring in this proof and so the exponents must satisfy this condition 1.) first – $k_1 - k_n$ 2. each derived from some preceding one where derived means

---

This property again purely arithmetic *proof number*

Further we consider the relation $\underline{yPrx}$

means x proof and last exponent of $x = y$

---

and class $P(x) \equiv (Ey)\, yPrx$

Since the notions P | and Pr | are arithmetic and as arithmetic contained in our system they can be expressed by formulas in fact by propositional functions as can be shown in detail.

So consider from now on P | and Pr | as abbreviations for complicated formulas which can actually be found and written

15.

Relation $yPrx$ constructive (finite number of steps) and this has the consequence

If A arbitrary formula and a its number if A provable then $P(a)$ provable

Proof: Suppose A provable and b number then $bPra$ true and provable hence $(Ey)\, yPra = P(a)$ provable

---

I need one more arithmetic notion derived from metamathematics $S(x, y)$ (*calculable*)

Again can be shown to be arithmetic notion which can be calculated (how?)

Therefore represented by a formula of our system and again consider S as denoting this formula

---

through with preparations

16.

Consider this expression

$$\sim P[S(x,x)] \quad \ldots q$$

this is a propositional function with one variable and means:

1. Formula obtained by substitution x in formula number x is not provable

2. The property expressed by propositional function number x cannot be proved to belong to the number x - computed

The above proposition f being a formula of our system it must have a number q (*calculable!*). Substituting q I get a proposition

$$\sim P[S(q,q)] \qquad\qquad S(q,q)$$

which says that proposition number $S(q,q)$ is not provable. What is the number of this formula $S(q,q)$?

Let's introduce r for $S(q,q)$ then

17.

$$\underbrace{\sim P(a)}_{A} \ \ldots \ a \quad \text{number}$$

A states on itself that it is not provable or arithmetic statement equivalent to statement A not provable

---

Now we prove

If A provable then system contradictory

Apply auxiliary theorem we have a proposition A with number a and know if A provable then $P(a)$ provable so

If $\sim P(a)$ provable then $P(a)$ provable

If $\sim P(a)$ provable system contradictory

If system consistent A not provable. But owing to the fact that A itself means exactly that A is not provable we may say

If system consistent then A, i.e.,

$$C \to A$$

if C means the statement that the system [is consistent],

18.

is not provable.

This statement can itself be expressed by formula owing to correspondence. So we proved a certain formula of our system $C \to A$ and this proof can be formalized so we have

$C \to A$ is provable

Now it follows that C cannot be proved because if it were provable then A were provable and then system contradictory.

[Cancelled: Close consideration shows constructivity]

19.

So we have shown: If the statement that our system is free from contradiction could be proved then our system would be contradictory and a closer examination shows that we could actually exhibit this contradiction, i.e., given a proof for freedom from contradiction we could derive from it an actual contradiction of our system. The second half of our program, the proof for the existence of undecidable arithmetic propositions is now easily accomplished. A, e.g., is such an undecidable proposition. For we know if our system is free from contradiction then A is not provable. [The last four lines have been lightly cancelled, then, it seems, the cancellation erased.] On the other hand we know under the same assumption that A is true therefore $\sim A$ is false and therefore $\sim A$ cannot be provable if we assume that no false arithmetic statement is provable in our system.

19.1

The proposition A which we proved to be undecidable is an arithmetic statement because P and S of which it is constructed are arithmetic notions.

But this proposition A seems at first sight to be very artificial and far remote from everything that is actually dealt with in arithmetic. This however is a wrong appearance. It can be shown that A can be transformed into a statement on the solutions of a certain diophantine equation, i.e., into a statement of the same character as are actually dealt with in number theory.

20.

Of course the undecidability of A is only relative. We can add a new axiom to one system which has the consequence that A becomes decidable in fact a very plausible axiom namely C which states that our system is free from contradiction. If we add this C, then owing to this implication A becomes provable but it would be wrong to suppose that now we should have obtained a system in which every arithmetic statement is decidable. For we can apply the same method of proof to our new system and construct another proposition, which is undecidable in the new system [added: For this system again we can add a new axiom, e.g., the statement that the new system is [consistent?] and make by it the undecidable proposition a decidable one], and so we can go on indefinitely with our ever reaching a system in which every arithmetic statement is decidable. This situation can also be expressed by saying: It is impossible to give a complete system of axioms for the arithmetic of integers, i.e., a system

21.

which makes it possible to decide any given arithmetic statement expressible in the primitive terms of our system. I wish to make a final remark on the impossibility of proving consistency. In that case too our statement is only relative, i.e., we proved only that if a definite formalization of mathematics is given then it is impossible to prove consistency of that formal system, i.e., using only the axioms and rules of inference of this same system. Someone may set up another formalism of mathematics and prove the consistency of the first system by an argument proceeding according to the rules of the second system. But we know in a proof for consistency the point is that it should be conducted by finite methods and now nobody has ever been able to produce a proof conducted by finite methods which could not easily be expressed in anyone of the formal systems for mathematics and nobody knows how to construct such a proof and therefore the foregoing considerations make it appear entirely hopeless to prove consistency for the transfinite methods of mathematics using only the unobjectionable methods of finite arithmetic which was the program of the formalistic school.

Can mathematics be proved consistent?

1.

The modern investigations in the foundations of mathematics gave as one of their remaining results the fact that all logic and all mathematics (at least as far as it has been developed so far) can be deduced by means of a few axioms and rules of inference. In order to bring out this fact clearly it was necessary at first to replace the imprecise and often ambiguous[1] by a perfectly precise artificial language, the logistic formalism. This formalism consists of a few primitive symbols representing the primitive notions of logic and mathematics [addition in Supplement I, cancelled: and the primitive symbols play the same role as the words in ordinary language]. I wrote some examples of those primitive symbols on the blackboard

$\sim$ not $\rightarrow$ implies *E* there exists

Now any logical or mathematical proposition can be expressed by a formula composed of these primitive symbols [addition in supplement I: one could even [cancelled: construct] device a machine that would give you as many consequences of the axioms as you like. The only trouble would be that it would give the consequences at random and therefore not the results one is interested in] and vice versa any formula composed of our primitive terms according to certain rules which constitute the grammar of our logistic language

2.

expresses (in a perfectly unique way) a definite mathematical statement.

[Addition from supplement II begins] In practice it would be very inconvenient to express mathematical statements in this way by means of the primitive terms of logic and mathematics. Our formulas would soon become very long and cumbersome. Therefore we usually introduce besides our primitive terms new symbols by definition but this process serves merely the practical purpose of abbreviation and from the theoretical point of view is entirely dispensible, as we can replace in any formula the new symbols by their meanings expressed on our primitive terms. Therefore we

[1] [The empty space left should be for an English equivalent of the German *Wortsprache* that Gödel used in his other presentations in the same context. The New York lecture has: colloquial language]

may look away entirely from the possibility of introducing new symbols by definition and think of any mathematical proposition as expressed by the primitive terms of logic and mathematics. The process of deduction is represented in this logistic language in the following way [text from supplement II ends]: Some of our formulas are considered as axioms, that is to say as the starting point for developing mathematics, and in addition to that certain rules of inference are stated which allow you to pass from the axioms to new formulas and this to deduce more and more propositions by iterated application of the rules of inference. One of the rules of inference, e.g., reads: If A and B are two arbitrary formulas and if you have proved the formula A and the formula $A \to B$, then you are entitled to conclude B [Cancelled: This rule of inference has an important characteristic An important feature of the rules of inference is this that they do not refer to the meaning of the formulas involved but only to their outward structure.

3.

You can see at once that in order to apply this rule of inference you need not know anything about the meaning of our formulas.]

The other rules of inference are of a similar simple character. In particular, all of them are purely formal, i.e., they do not refer to the meaning of the formulas but only to their outward structure so that they could be applied by someone who knew nothing about the meaning [cancelled: or by a machine]. By iterated application of our rules of inference, starting from our axioms, we obtain what I call [a derivation],[2] i.e. a [derivation] simply is a finite sequence of formulas $A_1 \ldots A_n$ which begins with some of our axioms and has the property that each of it's subsequent formulas can be obtained from some of the preceding one's by application of one of the rules of inference. Instead of [derivation] I shall also use the term formal proof or proof, and I will call a [derivation] a proof for formula F if the [derivation] under consideration ends up with the formula F. And of course we

4.

will call a formula F provable if there is a proof for it which means the same thing as: if F can be obtained from the axioms by iterated application of the

[2] [As explained in Part II, Section 4, Gödel had left space for an English word where he likely thought of the German *Herleitung*.]

rules of inference[3] A symbolism for which axioms and rules of inference are specified in the manner I have just described, is called a formal system[4] and one has succeeded in reducing all mathematics and logic to such a formal system. Owing to this fact, certain general questions concerning the structure of mathematics and logic have become amenable to a scientific treatment, whereas formerly those questions had to be left to vague speculations [added in supplement: and could not even be stated precisely]. I want to deal with two of these questions to-night. The first concerns the freedom from contradiction of mathematics. This question can now be stated in a perfectly precise

5.

way as follows: Does there exist any formula A such that A and $\sim(A)$ are both provable?, where the term "provable" has the precise meaning which I explained before, namely it means "being the last formula of a [derivation] which proceeds according to the rules of inference starting from axioms."

[Cancelled: The question of consistency is actually a mathematical question amenable to mathematical treatment. Every mathematician is convinced that there do not exist two such formulas A and $\sim A$ both provable, and so the problem of giving a proof for this statement.]

It can easily be shown that if two such formulas A and $\sim A$ which were both provable existed, then any formula whatsoever would be provable, for instance also the formula $0 = 1$, and so it is of vital importance for our formal system that this should not happen and the problem of giving a proof that it cannot happen arises. But at the same time, serious objections can be brought against this problem, namely one may, say, suppose we had found a proof for consistency, then owing to the fact that it is a mathematical proof, it must necessarily

[3] Cancelled addition on frames 411–412: It may seem as if in mathematics besides showing conclusions we would perform another important operation namely making definitions, i.e., introducing new notions by definition. But a closer consideration shows that definitions are by no means indispensable but merely serve for the sake of abbreviation as they introduce single symbols for complicated expressions. If, however, we take the trouble of allways writing down the complicated expressions in full we need no definitions at all and we can express every mathematical statement by means of the primitive notions of logic and mathematics alone.

[4] [Cancelled passage begins: and so I can state the fact to which I was referring in the beginning of my talk   has been established toward the end of the 19th century can be stated by saying that there are formal systems comprising all of mathematics now existing]

6.

proceed according to the axioms and rules of inference for mathematics. So in order that this supposed proof may prove anything at all we must know that our axioms and rules of inference which we used in it are correct, i.e., we must know that our axioms and rules of inference allways lead to correct results.[5] But if we know [changed from: believe] this in advance, then no proof for freedom of contradiction would be necessary, for A and $\sim A$ then could not both be provable, because they cannot both be correct results. Fortunately the actual situation is slightly different. For mathematics consists of two distinct parts which are usually referred to as finite and transfinite mathematics which may roughly be characterised as follows. Under the first heading are comprised (subsummed) all such considerations and methods of proof which do not presuppose the existence of any infinite sets and are based on this assumption, e.g., [cancelled: Let P be a property]

7.

[6]Now nobody has ever doubted [written above: questioned (challenged)] seriously the consistency of finite mathematics, whereas the situation is quite different with the transfinite methods based on the assumption of the existence of infinite sets which, by the way, is by far the greater part of now existing mathematics. In this domain of mathematics, actual contradiction had arisen unexpectedly toward the end of the 19th century, the so called paradoxes of the theory of aggregates. In order to avoid those paradoxes, certain restrictions on the previous assumptions concerning

[5] [Originally: So in order to be convinced by this supposed proof we must believe in the correctness of our axioms and rules of inference, i.e., we must believe that our axioms and rules of inference allways lead to correct results.]

[6] [Cancelled:

Let us call an integer a "prime sum" if it is the sum of two prime numbers. Now if we make the following association, "Either every even integer is a prime sum or there is an even integer which is not a prime sum," then the correctness of this assertion presupposes that the infinite totality of all even numbers exists somehow objectively because we assert the above disjunction although we are]

8.

infinite sets had to be made, and these restrictions[7] can be made in a very natural way and they do not alter anything in the mathematical results, but nevertheless, the faith of the mathematicians in the transfinite methods was shaken by the bad experience of the paradoxes of the theory of aggregates and there remains the fear that other paradoxes may arise in spite of our restrictions. And now you see what the question of proving consistency really is about. It is the problem of proving the freedom from contradiction of the transfinite mathematics by means of finite methods, i.e., using in the proof for consistency only such methods

9.

as are not based on the existence of infinite sets. So much on the meaning of the first problem, the question of consistency. The second problem which I want to deal with is so closely related to the first in its treatment that it can hardly be separated from it. It is the question of completeness of the formal system for mathematics, i.e., the question whether every mathematical statement expressed by a formula of our system can be decided by means of our axioms and rules of inference that is to say: Or in other words, is it true that if A is an arbitrary formula then either A or $\sim A$ is provable. I am going to sketch a proof which answers both those questions in the negative. I.e. 1.) It is not possible to prove our system consistent using only the finite methods of proof embodied in its axioms and rules of inference. 2.) There are propositions, in fact even

10.

arithmetic propositions A, for which neither A nor $\sim A$ is provable.

Of course mathematics can be formalized in different ways, that is to say, the axioms and rules of the formal system representing mathematics can be chosen in different manners and so one may suspect that our two results depend on the special system for mathematics which we chose. But that is not the case: It can be shown that the two theorems which I just stated hold good whatever formal system we may choose provided only that ordinary arithmetic of integers in its usual form is contained in our system

[7] [Cancelled: are of course carefully observed in the axioms and rules of inference of the formal system for mathematics, so that we may be quite sure that none of the known paradoxes of the theory of aggregates can be obtained by our axioms but of]

and that no false arithmetic statement is provable, and these two requirements seem to be indispensable for any system which can be claimed at all to represent mathematics.

The proof for these two statements (impossibility of a proof for consistency and existence of undecidable

11.

propositions is very cumbersome if worked out in all details, but I hope to succeed in (explaining you) making the leading ideas clear to you.

So let us suppose a formal system for mathematics given [cancelled: and let $f_1 f_2 \ldots f_k$ be its primitive symbols. Any formula then is a combination of those primitive symbols, i.e., a finite sequence of the primitive symbols.] Among the expressions which can be built up by our primitive symbols, also expressions of such a kind as I wrote down here occur:

$$x^2 > 6 \quad x^2 > y + 1 \quad x + y$$

$$\left\{ \begin{array}{ll} 2^2 > 6 & 2^2 > 2 + 1 \\ 3^2 > 6 & 2^2 > 5 + 1 \end{array} \right\}$$

Take, e.g., $x > 6$, it does not denote a proposition, but it becomes a proposition if the variable x is replaced by a definite number, e.g., x . 2.

Such an expression involving a variable which becomes a proposition if the variable is replaced by a number ist called propositional function. And we may say that propositional functions represent properties, e.g., $x > 6$ represents the property of being greater than 6 (or we may say they represent classes f_i). In a similar way, you can

12.

form propositional functions involving several variables which become propositions if [the variables are replaced by numbers], e.g., $[x > y]$.[8] Such propositional functions with several variables of course represent relations, e.g., [the relation of greater than]. Finally, there are expressions involving variables [cancelled: of a third kind (example) [possibly $x + y$ intended] which, if variables are replaced by numbers, do not become propositions, but become a symbol of definite numbers]. If here the variables are replaced you obtain a symbol for a definite number, so we have to bear in mind

[8] [Guessed from context.]

that these three kinds of expressions exist beside the expressions denoting propositions. That is first remark.

Suppose $f_1 f_2 \dots f_k$ be the primitive symbols of our system. Then every formula is a combination of these symbols = finite sequence of these symbols. [Cancelled: Therefore it is possible to order all the formulas in a series in a similar way as the words in dictionary are ordered. The only difference is that there are infinitely many formulas and therefore our dictionary would have – –] Owing to this fact, it is possible to number the formulas, i.e., to associate with each formula a number in the same way as to each house

13.

in a street a number is associated, of course in such a way that to different formulas different numbers. This numbering of the formulas can be effected in many way's. For the subsequent purposes, it is most convenient to do it in the following way: We give [changed from: associate] at first a number to each of our primitive symbols $f_i - - i$. As each formula is a finite sequence of [primitive symbols], to each formula there corresponds a sequence of integers $k_1 k_2 \dots k_s$, and from this sequence of integers, we construct a single integer in the following way: $2^{k_1} \cdot 3^{k_2} \cdot \ldots \cdot p_s^{k_s}$ [added above: and consider this as the number of] ? So to every formula a corresponding integer.[9] As we know, proof = sequence of formulas satisfying certain conditions. As any formula has a number, a proof can be represented by a sequence of numbers, and in the same way as before, we can again construct a single number out of this sequence and consider this as the number

14.

of our proof. So we have succeeded in numbering not only the formulas, but also the proofs.[10] Owing to this numbering of the formulas, of course to every class of formulas there corresponds a certain class of integers and to every relation between formulas a certain relation between integers. Take, e.g., the following relation: the formula x is longer than the formula y,

[9] of course not all numbers used up

[10] Not all numbers are used up in this numbering process, that is to say, there are numbers that correspond to no formula and numbers that correspond to no proof. But that does not matter; the only thing we need is that every formula and every proof should be labeled by a definite number.

i.e., consists of more symbols. What does that mean for the numbers of the formulas x, y which I denote by m, n? Of course it means that m contains more different prime factors than y. So we have a definite arithmetic relation which corresponds to the relation of being longer between formulas. In a similar manner to any

15.

relation between formulas, there corresponds a certain arithmetic relation between integers. Relations between formulas are called meta and so we can say – –. The situation is similar to that in analytic geometry where we also associate numbers (called coordinates) to certain entities namely the points, planes, etc., and owing to this correspondence to any geometric relation [continued in supplement IV] between points corresponds an arithmetic relation between numbers, for instance the relation which holds for three points if they lie on a straight line corresponds to the arithmetic relation of linear dependence between numbers. Similarly also to each geometric statement corresponds an arithmetic statement, e.g., and in our case to each metamathematical statement concerning the formulas of our system corresponds a certain arithmetical statement. [end of continuation]

I hope this will become clear by some other example which I need for the subsequent proof. Take, e.g., the relation of immediate consequence between formulas. What arithmetic relation does correspond to it? I called a formula R an immediate consequence of two formulas P and Q if R could be concluded from P and Q according to our rule of inference, which means the same thing as P must be an implication whose first term is Q and second term is R, because only when P has this structure we can apply our rule of inference and get R as a consequence. Now what does that mean for the numbers of P, Q, R ? According to our definition, if p corresponds to the formula P, then the series of exponents in the expansion of p corresponds to the series of symbols of which p [should be P] consists, and similarly for q and r. Now suppose the sign of implication which is one of the primitive terms has the number k.

16.

Then the fact that the formula P is composed of the formulas Q and R with an implication sign between them means for the corresponding number that the series of exponents in the expansion of p is composed with the number k between them. This is clearly a purely arithmetic relation between the numbers p, q, r, and if three numbers are given, it can easily be ascertained whether this relation holds or not. And if and only if this arithmetic relation holds between three integers p, q, r, the relation of immediate consequence holds for them. [Addition in supplement IV] If three integers p, q, r stand in this relation, I shall call r derived of the two integers p and q. Now we are in a position to give a purely arithmetic characterisation of those integers which correspond to proofs [end of addition]. Also the class of integers which are numbers of some proofs [cancelled: can be characterised arithmetically] in the following manner. Let us suppose that we have n axioms and that the number of these n

17.

formulas taken as axioms be $k_1 k_2 \ldots k_n$. So there are n definite numbers which are determined if the system is given.[11] A formal proof of our system was defined as being a sequence of formulas beginning with some axioms and such that each subsequent formula of the sequence can be obtained from some of the preceding ones by the rule of inference. What does that mean for the corresponding number, i.e., what arithmetic property must a number a posess in order to be the number of a proof? Now according to the manner in which we associated numbers to proofs we have this: If a corresponds to a proof, then the exponents in the prime expansion of a must correspond to the single formulas of this proof, and so these exponents must satisfy the condition that the first exponents say m_1 up to m_s, must be some of the n numbers $k_1 \ldots k_n$, and each of the subsequent exponents must be derived from some of the preceding, where derived means that purely arithmetic relation which I defined before, so the property of being the number of some proof can be expressed purely arithmetically. Therefore also this relation P

18.

which means [$y \, Pr \, x$] can be expressed arithmetically, because it simply means y is the number of a proof and the last exponent of y is equal to

[11] Let us further suppose, for the sake of simplicity, that we have only one rule of inference. This one which [perhaps intended: with] implication. Actually there are several rules of inference but that makes no essential difference.

x. Further, we can form the class of those integers which correspond to provable formulas and denote it by P; so $P(x)$ means the formula with the number x is provable, i.e., $(Ey)y\,Pr\,x$

[Addition from supplement III] This property P again is a purely arithmetic property. Now we have assumed that arithmetic be contained in our formal system, that is to say that arithmetic relations and statements can be. Therefore, relation Pr and the property P, being purely arithmetical, can be expressed by formulas of our system. We know how relations and properties are expressed in a formal system. They are expressed by propositional functionals, so we can say, and it can be shown with all necessary rigor, that there is a propositional function with two variables [continues in supplement IV] expressing the relation $y\,Pr\,x$, function $P(x)$ with one variable expressing this property, and it would make no difficulty actually to write down these formulas as in terms of the primitive symbols of the system of mathematics and logic, and so I beg [above: ask] you to consider these letters P, Pr from now on as a shorthand notation for some long and complicated formulas of our system [end of addition].

Now the relation $y\,Pr\,x$ has the following important property: If two arbitrary numbers k, l are given, you can allway decide in a finite number of steps whether for these two numbers the relation Pr holds, i.e., whether k is [the number of a proof for the formula number] l. Because in order to find that out you have only to expand k in its prime factors (which can – – step), then write down the series of exponents and find out 1.) if the last exponent is the number l 2.) if the first exponents up to a certain stage are numbers of axioms and 3.) If all the subsequent numbers are derived from the preceding one's. And as the number of exponents for which you have to check the latter statement is finite and as the relation of "being

19.

derived" is [decidable], this can allways be done in a finite number of steps. This fact that for any two integers k, l it can be decided in a finite number of steps whether the relation Pr holds or not now has the following consequence: Let A be an arbitrary formula and a its number. Then I assert if A is provable $P(a)$ is provable, i.e., if A is a provable formula, then for the number of a one can prove that it is the number of a provable formula. So this theorem which I assert means: If A is provable, then the statement that A is provable is also provable. [Addition in supplement III] I.e., if the statement that A is provable is true it is also provable. It could not be asserted for every mathematical proposition that if it was true, it was provable, but it can be asserted for a certain class of proposition (existence propositions). Take for instance the famous theorem of Goldbach which states that [every even integer is a] prime [sum]. If it was false we could certainly prove it was false [end of addition].

The reason for this theorem's holding is this: Suppose A is provable, i.e., there is a proof. This proof has a number, say b. b is the number [of a proof for the formula] a, i.e., $b\, Pr\, a$, but as Pr is decidable for any two numbers, we can ascertain in a finite number of steps that $b\, Pr\, a$, i.e., prove $b\, Pr\, a$, and therefore also prove $(Ey)y\, Pr\, a$, i.e., $P(a)$.

20.

For the subsequent considerations, I need one more arithmetic notion derived from a corresponding metamathematical, and this is the following [from formula list]:

$S(x, y)$ = number of formula obtained from formula number x by substituting number y for the variable a

So if two numbers x, y are given, $S(x, y)$ allways is a definite number which can be computed in a finite number of steps. In order to do that we have first to find out whether the combination of symbols corresponding to x is a propositional function. If this is not the case, we know that $S(x, y) = 0$, if it is the case, we have to replace the variable in this propositional function by the number y. Thus we obtain a definite formula and we have only to compute the number of this formula. This gives us the value of $S(x, y)$. Again, this function $S(x, y)$ whose arguments as well as its value are integers turns out to be a purely arithmetical function which therefore can be represented by a formula of our system, and again consider the latter S as a shorthand notation for this formula. Now I am through with the preparations and can

21.

proceed to the proof. I consider the following expression:

$$\sim P[S(x, x)]$$

This is a propositional function with one variable x and it means: the formula obtained from the propositional function number x by substituting in it the number x itself is not provable. If x is not the number of a propositional function, then this formula is allways true, because then $S(x,x) = 0$. But that is of no importance. If x is the number of a propositional function, then this statement means the following thing: the property expressed by the propositional function number x cannot be proved to belong to the number x. Now this formula being itself a propositional function of formal mathematics, it must have a number itself. Let us call it q. This number could easily be calculated if we like because – – . And now I substitute the number q in it so that I obtain [$\sim P[S(q,q)]$]. So we get a definite proposition which says that

22.

the proposition number $S(q,q)$ is not provable, i.e., that the proposition obtained [$\sim P[S(q,q)]$] is not provable. What is the number of this proposition? We can easily compute it from the manner in which we obtained it; We obtained it by substituting the number q in this propositional function, i.e., in the propositional function with the number q. But the number of the formula obtained from [formula number q] by [substituting number q for the variable x] is $S(q,q)$. That is exactly the definition of S so we know that this proposition, I call it R, has the number $S(q,q)$. [Added: Lets introduce an abbreviation for this expression $S(q,q)$, say r, then we have $\underbrace{\sim P(r)}_{R}$, [formula number] r, and I call this statement R.] But what does this proposition state? It states that proposition N$^{\underline{o}}$ r is not provable, but the proposition number r is exactly itself. So we have a proposition which states about itself that it is not provable. This proposition really is an arithmetical statement because P and S are arithmetical functions and so to be more exact we would have to say we have an arithmetical statement

23.

R which is equivalent with the logical statement that R is not provable. And now the next step is that we can prove that if this proposition were provable, our system would be contradictory. For this purpose, we have only to apply the above theorem that if r is the number of R and if R is provable, then the formula $P(r)$ is provable. So in our case, if the proposition R number r is provable, then the formula $P(r)$ is provable,

24.

i.e., if $\sim P(r)$ were provable, then system for mathematics would involve contradiction, or vice versa if system consistent, then R not provable. But R just means that R is not provable and so we may say – –

Now owing to the fact that we have replaced the notions relating to formulas (e.g. notion of provability) by arithmetic notions, also to every metamathematical statement relating to formulas there corresponds an equivalent arithmetic statement, e.g., also to the statement that our system is consistent. I denote this statement by *Contr.* So what we have proved so far can be understood to be an arithmetical statement. So this arithmetical statement is proved and therefore for this also, a formal proof for it in our system can be given, in our system. But from this we infer at once that C cannot be proved. For if it could be proved, then owing to this implication also R could be proved but if R could be proved, the system would be contradictory.

25-415 [The three formula pages follow]

1.)]

$\sim$ not $\quad\rightarrow$ implies $\quad E$ there exists

$\sim p \quad p \rightarrow q \quad (Ex)F(x)$

$\left.\begin{array}{c} A \rightarrow B \\ A \\ \hline B \end{array}\right\}$ premises $\quad$ Rule of inference

$A_1\,A_2 \ldots A_n$ chain of inference or formal proof

$x^2 > 6,\ x > y^2 + 2$ propositional functions

$\sim,\ \rightarrow,\ E,\ x,\ s_5,\ s_6 \ldots s_n$

$1,\ 2,\ 3,\ 4,\ 5,\ \ldots\ \ n$

$A \ldots n_1\,n_2 \ldots n_r \ldots 2^{n_1}3^{n_2} \ldots p_r^{n_r}$

$p_r =$ *r-th* prime number

$\left.\begin{array}{c} A_1\,A_2\ \ A_m \\ a_1\,a_2\ \ a_m \end{array}\right\} \quad 2^{a_1}3^{a_2} \ldots p_m^{a_m}$

$$\begin{array}{cc} P & Q \to R \\ \underline{Q} & \underline{Q} \\ R & R \end{array} \quad \Bigg| \quad P = Q \to R \quad \begin{array}{l} p = 2^{a_1} \ldots p_l^{a_l} \\ q = 2^{b_1} \ldots p_m^{b_m} \\ r = 2^{c_1} \ldots p_n^{c_n} \end{array}$$

$(a_1, a_2 \ldots a_l) = (b_1, b_2 \ldots b_m, 2, c_1, c_2 \ldots c_n)$
r derived from p, q

25-416

2.)⌋

$k_1\, k_2 \ldots k_n$ numbers of axioms

$z = 2^{m_1} \ldots p_r^{m_r}$

$m_1 \ldots m_s \mid m_{s+1} \ldots m_r$

$y\, Pr\, x = y$ is the number of a proof for the formula number x

---

$P(x) =$ formula number x is provable = $(Ey)\, yPr\, x$

$S(x, y) =$ number of formula obtained from formula number x by substituting number y for the variable a

---

A number a

If A provable $P(a)$ provable

25-417

3.)⌋

$\sim P[S(a, a)]$ number q

$\sim P[S(qq)]$ number $S(qq)$

$$S(qq) = a$$

$\underbrace{\sim P(a)}_{A}$ number a

If $\sim P(a)$ provable then $P(a)$ provable

If $\sim P(a)$ provable system contradictory

If system free from contradiction A not provable

If system free from contradiction A holds

$C \to A$ is provable

C = statement that system free from contradiction

Index of names in Gödel's notes:

References in Gödel's notes:

Gödel's shorthand notes and the typewritten manuscripts contain, next to references to the *Principia Mathematica* in continuation, explicit references to only the following sources:

Bernays, P. (1930) Philosophie der Mathematik und die hilbertsche Beweistheorie. *Blätter für Deutsche Philosophie*, vol. 4, pp. 326–367: 278L

Fraenkel, A. (1927) *Zehn Vorlesungen über die Grundlegung der Mengenlehre*: TM 1

Gödel, K. (1930) Über die Vollständigkeit der Axiome des logischen Funktionenkalküls. *Monatshefte für Mathematik und Physik*, vol. 37, pp. 349–360: TM 37

Gödel, K. (1930) Einige metamathematische Resultate über Entscheidungsdefinitheit und Widerspruchsfreiheit. *Anzeiger der Akademie der Wissenschaften zu Wien, Mathematisch-naturwissenschafliche Klasse*, vol. 67, pp. 214–215: TM 1

Hilbert, D. (1928) Probleme der Grundlegung der Mathematik. *Mathematische An-*

J. von Plato, *Can Mathematics Be Proved Consistent?*, Sources and Studies in the History of Mathematics and Physical Sciences, https://doi.org/10.1007/978-3-030-50876-0

nalen, vol. 102 (1929), pp. 1–9: TM 32

Hilbert, D. and W. Ackermann (1928) *Grundzüge der theoretischen Logik*: 283L, TM 36, TM 37

von Neumann, J. (1927) Zur Hilbertschen Beweistheorie. *Mathematische Zeitschrift*, vol. 26, pp. 1–46: 274R, TM 11

von Neumann, J. (1928) Die Axiomatisierung der Mengenlehre. *Mathematische Zeitschrift*, vol. 27, 1928, pp. 669–752: TM 1, TM 32

REFERENCES FOR PARTS I AND II:

Bernays, P. (1918) *Beiträge zur axiomatischen Behandlung des Logik-Kalküls*. Manuscript Hs. 973:193, Bernays collection, ETH-Zurich. Printed in Hilbert (2013).

Carnap, R. (1929) *Abriss der Logistik*. Springer, Vienna.

Carnap, R. (1931) Die logizistische Grundlegung der Mathematik. *Erkenntnis*, vol. 2, pp. 91–105.

Davis, M. (1965) *The Undecidable. Basic Papers on Undecidable Propositions, Unsolvable Problems and Computable Functions.* As republished by Dover, 1993.

Dawson, J. (1997) *Logical Dilemmas: The Life and Work of Kurt Gödel*. A. K. Peters.

Fraenkel, A. (1928) *Einleitung in die Mengenlehre*. 2nd ed., Springer.

Goldfarb, W. (2005) On Gödel's way in: the influence of Carnap. *The Bulletin of Symbolic Logic*, vol. 11, pp. 185–193.

Hempel, C. (2000) An intellectual autobiography. In *Science, Explanation, and Rationality*, ed. J. Fetzer, pp. 3–35. Springer

Herbrand, J. (1930) *Recherches sur la théorie de la démonstration*. Thèses présentées à la faculté des sciences de Paris, 128 p.

Heyting, A. (1931) Die intuitionistische Grundlegung der Mathematik. *Erkenntnis*, vol. 2, pp. 106–115.

Hilbert, D. (1899) *Grundlagen der Geometrie*. Teubner, Leipzig.

Hilbert, D. (1900) Mathematische Probleme. *Nachrichten der Königlichen Gesellschaft der Wissenschaften zu Göttingen, mathematisch-physikalische Klasse*, pp. 253–297.

Hilbert, D. (1918) Axiomatisches Denken. *Mathematische Annalen*, vol. 78, pp. 405–415.

Hilbert, D. (1928) Probleme der Grundlegung der Mathematik. As reprinted in *Mathematische Annalen*, vol. 102 (1929), pp. 1–9.

Hilbert, D. (2013) *David Hilbert's Lectures on the Foundations of Arithmetic and Logic*,

1917–1933. Edited by W. Ewald and W. Sieg. Springer.

Hilbert, D. and W. Ackermann (1928) *Grundzüge der theoretischen Logik*. Springer.

Mancosu, P. (1999) Between Vienna and Berlin: The immediate reception of Gödel's incompleteness theorems. *History and Philosophy of Logic*, vol. 20, pp. 33–45.

Mostowski, A. (1952) *Sentences Undecidable in Formalized Arithmetic: An Exposition of the Theory of Kurt Gödel*. North-Holland.

von Neumann, J. (1927) Zur Hilbertschen Beweistheorie. *Mathematische Zeitschrift*, vol. 26, pp. 1–46.

von Neumann, J. (1931) Die formalistische Grundlegung der Mathematik. *Erkenntnis*, vol. 2, pp. 116–121.

von Plato, J. (2017) *The Great Formal Machinery Works: Theories of Deduction and Computation at the Origins of the Digital Age*. Princeton University Press.

von Plato, J. (2018a) In search of the sources of incompleteness. In *Proceedings of the International Congress of Mathematicians 2018*, vol. 3, pp. 4043–4060.

von Plato, J. (2018b) Kurt Gödel's first steps in logic: formal proofs in arithmetic and set theory through a system of natural deduction. *The Bulletin of Symbolic Logic*, vol. 24, pp. 319–335.

Rand, R. (2002) Wechselrede zum Referat Herrn Gödels. In E. Köhler et al., eds, *Kurt Gödel: Wahrheit & Beweisbarkeit*, vol. 1, pp. 133–134. öbv et hpt VerlagsgmbH & Co, Vienna.

Skolem, T. (1929) Über einige Grundlagenfragen der Mathematik. *Norsk videnskaps-akademi i Oslo. Skrifter I. Mat.-naturv. klasse*. No. 4.

Tarski, A. (1935) Der Wahrheitsbegriff in den formalisierten Sprachen. *Studia Philosophica*, vol. 1, pp. 261-405.

Wang, H. (1996) *A Logical Journey: from Gödel to Philosophy*. The MIT Press.

Whitehead, A. and B. Russell (1910–13) *Principia Mathematica*. Vols. I–III. Cambridge. Second edition 1927.

Zermelo, E. (1932) Über Stufen der Quantifikation und die Logik des Unendlichen. *Jahresbericht der Deutschen Mathematiker-Vereinigung*, vol. 41, pp. 85–88.

Printed in the United States
by Baker & Taylor Publisher Services